Plant IT: Integrating Information Technology into Automated Manufacturing

Part II: Integrating Information Technology into Automated Manufacturing

Plant IT: Integrating Information Technology into Automated Manufacturing

DENNIS L. BRANDL AND DONALD E. BRANDL

MOMENTUM PRESS

MOMENTUM PRESS, LLC, NEW YORK

Plant IT: Integrating Information Technology into Automated Manufacturing

First published in 2012 by Momentum Press®, LLC
222 East 46th Street, New York, NY 10017
www.momentumpress.net

ISBN-13: 978-1-60650-431-4 (hard cover, case bound)
ISBN-10: 1-60650-431-2 (hard cover, case bound)
ISBN-13: 978-1-60650-433-8 (e-book)
ISBN-10: 1-60650-433-9 (e-book)

DOI: 10.5643/9781606504338

Cover Design by Jonathan Pennell
Interior design by Exeter Premedia Services Private Ltd.,
Chennai, India

10 9 8 7 6 5 4 3 2 1

Printed in the United States of America.

Contents

Introduction

Information Technology is everywhere. It has become such a pervasive part of our life through instantly available information, instant communication, and social networking that we forget that these changes have occurred in a relatively short time. It is not surprising that these changes have not yet spread to all parts of businesses, because these changes are so new while some business processes change so slowly. Manufacturing, in particular, lags the general introduction of new IT systems and concepts. This lag is usually the result of significant capital investment in manufacturing equipment and technologies and their use for 10, 20, or 30 years before replacement. IT is slowly penetrating manufacturing operations as older equipment is upgraded or replaced and as new facilities are added. IT-based systems are also becoming more common as older precomputer engineers retire and newer computer-savvy engineers, who are looking for more IT solutions to manufacturing problems, replace them.

This book is based on concepts from the Manufacturing IT columns published in *Control Engineering* magazine. Each section discusses an IT issue that is important to a manufacturing company. It is assumed that the reader is somewhat familiar with manufacturing and what it takes to operate modern manufacturing equipment. In many cases, IT is related to automation and control, and it is also related to how manufacturing personnel obtain and use the information needed for their work. A list of acronyms at the end of the book will help you master IT terminology.

Our goal in this book is to educate manufacturing personnel and automation professionals in some of the concepts that have become increasingly important in creating highly productive manufacturing sites. It will not present everything you need to know, but will introduce you to the terms and concepts required for a modern control professional.

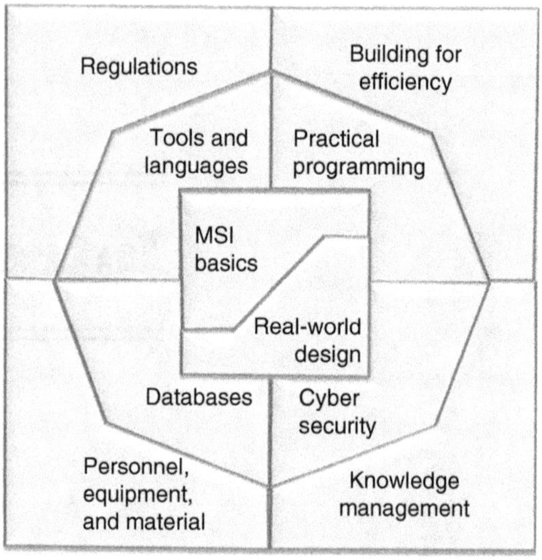

Figure 1. Plant floor IT.

Figure 1 helps visualize the layout of this book, a collection of different topics, that when combined provide a solid basis for covering many plant floor IT topics. Each chapter and section within each chapter can stand alone. If any specific area is important to your current job or project, you do not have to read the previous sections, but we encourage you to look at all of the sections, there may be information that will help you in unexpected ways. Note that tables and figures are numbered consecutively throughout the book, to aid in quickly locating them.

MSI (Manufacturing Science Informatics) and real-world design are the core elements of plant floor IT. These chapters cover some of the basic requirements for manufacturing professionals.

The basics are augmented with chapters on tools and languages used in plant flow systems, practical programming considerations for manufacturing software, databases, and security. The tools and software languages for manufacturing are unique and different from traditional IT tools and languages. Practical programming covers the differences between the typical manufacturing environment and typical business environment. Databases are a vital part of most manufacturing systems and are covered in a separate chapter. Cyber security is a growing component of manufacturing systems and is also covered in a separate chapter.

The outer ring of Plant Floor IT elements are regulations, building for efficiency, knowledge management, and information about the manufacturing resources such as personnel, equipment, and material. Regulations can be a critical element for some manufacturing companies, and a chapter covers some of the regulations that are important. Building for efficiency covers a range of IT-related topics, including service-oriented architectures (SOAs), computer network structures network, standards, and scheduling models. There are three main resources used in manufacturing: personnel, equipment, and materials. A chapter covers models that can be used to represent and exchange resource information. Knowledge management deals with preserving manufacturing knowledge in projects gained by older personnel older and consolidating it before they retire.

MSI Basics

Manufacturing science has been described as the body of scientific knowledge, regulations, and principles involved in the transformation of materials and information into products. Informatics is a broad term covering a wide range of IT elements that are important to companies. It includes sales, inventory, human resources, ERP, PLM, and other systems. Manufacturing Science Informatics (MSI) specifically relates to manufacturing and includes process control and automation systems, MES, and knowledge of analytical tools and techniques for collecting, storing, and analyzing manufacturing data.

MSI essentially comprises a checklist of interconnecting technologies about which a modern control professional must have some basic knowledge. This chapter presents an overview of MSI elements, several aspects of software engineering, and a set of organization attributes where different MSI activities take place. Subsequent chapters provide more in-depth information about the MSI elements and their typical organizational settings.

Software engineering is a foundation for all IT elements, yet many manufacturing engineers are tasked with software development when it is not their area of training. Software engineering is a structured approach to the design, development, operation, and maintenance of software and execution of software projects. Software projects are different from many traditional engineering projects and it is important to understand the software engineering methods discussed in this chapter. The chapter also includes some rules for successful software projects. These rules are empirical; having been developed from multiple successful and unsuccessful projects, and will help you start a new successful software project.

Manufacturing professionals understand that what you do not measure you cannot improve and so the chapter provides an introduction to two commonly used software measurements. The Software Engineering Institute

Capability and Maturity model and the Personal Software Process model provide measures that allow engineers to measure themselves and their organizations. These models consider different organizational resources and also provide methods for identifying improvements in productivity and software product quality.

WHAT'S IN A NAME?—MANUFACTURING SCIENCE INFORMATICS

Manufacturing engineering encompasses, among other specialized areas, knowledge of instrument engineering, control engineering, chemical engineering, mechanical engineering, and electrical engineering. When information technology is added it means that engineers must also have knowledge of informatics technologies and software engineering. Informatics is the application of information technology and statistical techniques to the management of information. Manufacturing Science Informatics (MSI) can be defined as the application of information technology to manufacturing. Table 1 provides a visualization of the MSI technologies and their categories and it should not be considered complete. It is a convenient schematic to understand what you need to know about MSI and we encourage everyone to extend it with other knowledge areas acquired in applying MSI.

Software engineering is the foundation for all of the elements of MSI. This includes understanding source code control, configuration management, defect tracking, software project life cycles, and software project management methods. Software engineering provides the conceptual structure for the individual elements that make up MSI. Traditional software engineering covers normal IT elements, but the addition of manufacturing IT requires additional elements for real-time control systems and the manufacturing IT-based standards and models.

IT basics include knowledge of object-oriented programming models and of high-level languages, such as Java, C++, C#, and Basic. While you do

Table 1. Manufacturing Science Informatics

Software Engineering			
IT Basics	Languages & Tools	Networking	Databases & Data Models
Real-Time Basics	IEC 61131-3 Control Languages	Control System Communication	Analytical Tools
Standards and Models	ISA 88 Modular Automation Models	ISA 99 Security ISA 84 Safety	ISA-95 MOM/MES Models

not need to be an expert in each language, you should know the strengths and weaknesses of each so you can apply the right tool to the task. It is also important to have some knowledge of other special purpose languages, such as PERL and PHP for any web-based development, XSLT for data transformations, or command line scripting languages such as Microsoft's PowerShell for efficient system management. IT basics also include knowledge of common IT tools, such as Enterprise Service Buses, cloud computing, virtualization, and knowledge management.

In addition to standard IT languages, a sound knowledge is required of the IEC 61131-3 programming languages (Ladder Diagrams, Function Block Diagrams, Instruction List, Structured Text, and Sequential Function Charts). IEC 61131-3 also implies a system architecture that is fairly common across PLC and DCS systems. These languages are designed to handle the real-time, interrupt-driven, guaranteed response time, fixed memory use, and limited CPU requirements of automation and control systems. The environment of the IEC 61131-3 languages, PLCs and DCSs, often provide redundancy, fast startup, intrinsic communications, and memory management, simplifying the implementation of control strategies without worrying about the underlying operating system.

A good knowledge of the IEC 61512 or ISA 88 Batch Control (www.isa.org) standard is needed because it provides a structure for well-designed 61131-3 code. This structure, usually called Modular Batch Automation, has been applied to all areas of control, including discrete, continuous, and batch control. The ISA 88 models define a structure for organizing the language elements of both the IEC 61131-3 languages and higher-level languages. It defines a modular structure designed for maximum reuse of code and maximum flexibility for system expansion and enhancements. There are multiple great books that explain the ISA 88 standard that are listed in the bibliography. The next set of MSI categories deal with communication and security. MSI requires basic computer science knowledge of networking and network services. This includes the best practices for network infrastructure using a hierarchy of switches, routers, and VLANs to isolate control networks from business networks and provide robust and resilient solutions. This also requires knowledge of network services such as domain name services (DNS), DHCP, NetBIOS, and network protocols such as TCP/IP and UDP/IP. Some knowledge is required of Access Control Lists (ACLs), Microsoft Windows Server tools, Active Directory Domains, and Linux tools.

Real-time basics extend networking knowledge to include device and control system communications. Today, this is largely based on the OPC standard (www.opcfoundation.org) and various fieldbus technologies, such as Foundation Fieldbus, Profibus, and DeviceNet. Industrial networks are designed for consistent performance, with deterministic response times, and are usually optimized for small data transfers corresponding to values from sensors or commands to control equipment. Newer device and control system networks are becoming increasingly based on Ethernet technologies, so there is starting to be considerable overlap with standards IT network solutions.

Cyber security is an increasing important element of manufacturing systems and understanding of the requirements and solutions is vital. The ISA 99 technical report and standards define network architectures for secure communications, which include Ethernet-based IT networks and industrial networks. In addition, the ISA 99 documents, and associated IEC versions, define the policies and procedures that must be put into place to design and maintain secure systems. Additional security constructs are available from British Center for the Protection of National Infrastructure (CPNI) at www.cpni.gov.uk

Safety has always been important in automation and control systems, and basic information is defined in the ISA 84 and IEC 61511 Safety Integrated Systems standards. These should be required reading for any manufacturing engineer. The bibliography lists several books that provide good information on industrial safety and the related standards.

One IT basic is knowledge about databases; in particular, a good knowledge of relational databases and SQL (ANSI/ISO 9075) is required. This should be complemented by knowledge of data models, such as Unified Modeling Language (UML), as defined in the ISO/IEC 19501-1 UML specification, XML (www.w3.org/XML) and XML Schema Definitions (XSDs). An MSI engineer should be able to create and read UML specifications; define, normalize, build, and query databases using SQL; read and understand XML and XSD documents; and know how all of these tools and models can be used in system development and integration.

Analytical tools are a common element of real-time systems, both for the analysis and for the optimization of physical processes, manual procedures, and automated control strategies. This knowledge augments IT basics for the tools necessary for increasing manufacturing productivity. This includes an understanding of real-time data historians, SPC for process control, SQC for quality control, standard statistical analysis techniques, Pareto Analysis, fish bone diagrams, six-sigma study techniques such as the Define-Measure-Analyze-Improve-Control (D-M-A-I-C) cycle, design of experiments, and multivariant analysis.

The last element is an understanding of the Manufacturing Operations Management (MOM) and Manufacturing Execution Systems (MES) models that are defined in the ISA 95 and IEC 62264 standards. These standards define both a structure and a definition of information that can be exchanged between plant floor systems and other business systems, and define a comprehensive model for all of the activities that occur in manufacturing operations management, including production, quality testing, inventory and material movement and management, and maintenance management. They provide a structure for understanding how personnel, equipment, and materials are used in manufacturing operations.

These brief descriptions define just some of the MSI knowledge required for manufacturing engineers. The information technology field is expanding, and we are continually being offered more opportunities for MSI to improve manufacturing and increase productivity.

TRUST IN SOFTWARE ENGINEERING METHODS

It seems that every generation needs to relearn the lessons of previous generations. This is equally true in manufacturing where older engineers are retiring and must pass on knowledge to younger engineers. Unfortunately, for manufacturing software, there are actually few lessons that can be passed on. Much of the knowledge required for robust manufacturing software development is not known to older engineers. The languages, tools, and methodologies for software development have been continually evolving and changing during their careers. Manufacturing software development lessons must come from outside manufacturing disciplines and should come from the software engineering discipline.

Many lessons of software engineering were acquired through the development of large systems that had lifetimes of dozens of years. This is a familiar situation in manufacturing, even as it is becoming increasingly rare in modern nonmanufacturing and nonregulated software systems. Many modern systems are designed for short lifetimes, 5 years at most before replacement. Many organizations also assume that the original programmers will often be available for maintenance and support. This is not the case for many manufacturing systems, which are installed and then not modified until the equipment must be replaced or significant new functionality is required. Fortunately, developers of manufacturing software can still learn from traditional software engineering rules.

One of the harsh truths about manufacturing software is that is it not developed by software engineers or computer science majors. We would not regularly ask an electrical engineer to design a mechanical system, or a chemical engineer to design an electrical system, but we often ask mechanical, electrical, and chemical engineers to design and develop software systems.

If you are in this situation, you should TRUST software engineering methods as defined below:

- T – Try-out prototypes
- R – Requirements
- U – Use cases
- S – Source code, standards, scripts
- T – Test cases, test scripts

Try out prototypes

The "T" represents try-outs and prototypes. Engineering disciplines use physical or mathematical models to understand the basic concepts and relationships of a system. Prototypes perform the same role for software. Prototypes can be throw away try-outs or incremental design elements, depending on your development methodology. However, they should be a common starting point for all projects.

Just because someone has a good concept for software project does not mean the project is doable, and hence a try-out confirms that the concept is technically feasible. By analogy, in automotive engineering, just because someone imagines a car that gets 300 MPG does not mean that it is feasible; prototypes prove or disprove the feasibility.

If you are in a project where you have not built or maintained a similar system, then try out different ideas and methods, and test the prototypes to ensure that the solution will actually work. This reduces the risk of missing hidden or assumed requirements and reduces the risk of major redesign late in the project.

Requirements

The "R" represents requirements. Once you know that the project is technically feasible, you can write the user, functional, and design requirements. These document what must be done and how the final system must operate in your manufacturing environment.

Many projects will skip the documentation of requirements, yet these documents provide information for testing, user documents, and support.

Use cases

The "U" represents use cases. While requirements documents are important, they do not document how the software is to be used by the end user. The use cases define normal operating methods and expected error conditions and responses. The use cases are designed to help the end user understand what will be provided.

Use cases can also be used to define the "stresses" the system must endure, such as the maximum number of users or maximum scan rate that must be achieved.

Source code, standards, and scripts

The "S" represents source code, standards, and build scripts. Source code should be developed only after you understand the requirements and use cases.

Source code can be developed in small chucks if you follow an agile project methodology or in large chunks if you follow an iterative or waterfall methodology.

Commenting, configuration management, and copyright standards must be applied to all generated code and documentation. Build scripts allow for daily or weekly automated recreation of the project software on a fresh system, eliminating manual errors and configuration issues.

Test cases and test scripts

The second "T" stands for test cases and test scripts. These are the detailed definitions that are used to verify that the software operates as specified under normal and error conditions. Software is not ready for deployment until it has successfully run the test cases. These are the final proofs of correctness.

Remember TRUST

TRUST in the lessons learned from software engineering and apply the TRUST model to your manufacturing IT projects. So remember to:

- **T** – Try-out prototypes
- **R** – Requirements
- **U** – Use cases
- **S** – Source code, standards, scripts
- **T** – Test cases, test scripts

RULES FOR SUCCESSFUL SOFTWARE PROJECTS

Many new information technologies can be effectively used in manufacturing environments, but it is difficult to keep up with all of the new choices and even harder to run effective projects that integrate the new technologies with existing systems. The sad fact is that several new technology manufacturing IT projects fail to meet their goals. However, there are some simple rules to apply to software projects that significantly reduce the risk of failure.

All software must:

1. Be fast and easy to implement and configure.
2. Not require a business process change.
3. Have no impact on existing software or hardware.
4. Use the existing IT security strategy and deployment methodology.
5. Have a low Total Cost of Ownership (TCO).

Fast and easy to implement and configure

Implementation and configuration should be measured in minutes and not hours or days. If the system is difficult to install or requires significant O/S and hardware expertise, then it indicates underlying problems. It may show immaturity of the software or the vendor, or it indicates a vendor's "we are so smart" attitude that says a user has to be an expert in the system just to get it up and running. Many manufacturing IT projects require controller hardware and I/O interfaces, so hardware installation can take significant time, but software should have self-discovery and configuration features to eliminate installation problems.

Do not require a business process change

Requiring business process changes for new software increases the installation, training, and reengineering effort. Required business process changes are usually the reason that ERP software is so difficult and expensive to install. Affected business process in manufacturing IT projects are usually configuration management, change management, inventory reporting, and business system integration. Systems that force business process changes instead of adapting to your processes may also be inflexible in other ways and should be avoided.

Have no impact on existing software or hardware

Changing the existing infrastructure will be expensive and time consuming. Typically, this problem is a result of an infrastructure mismatch.

Examples:

- The new system requires a Linux server, but you have standardized on Microsoft servers.
- The new system requires a Microsoft SQL, but you have an Oracle database standard.

Use the existing IT security strategy and deployment methodology

If the new software or system has its own security strategy or deployment methodology, then it will come into conflict with corporate standards. This means there will be extra time required to modify the software or modify corporate standards. In addition, because of continuing IT-based threats, many corporate IT security groups can veto any project considered unsafe or unreliable.

Have a low TCO

Every installed system carries a maintenance burden measured in Total Cost of Ownership. Programs that require manual updating of databases or manual copying of data with other programs will increase the maintenance burden. High "after first year" costs often result in systems that do not get needed support and attention and their value quickly degrades over time.

Break as few rules as possible

My estimate is that violating any single rule will cut your chance of success by at least a third or will double your development and deployment costs. It may be impossible to avoid breaking all of the rules because of hardware and control constraints in a manufacturing IT project, but when you have the choice, it is important to follow these simple rules for successful software projects. Remember to:

- Be fast
- No business process change
- No software or hardware impact
- Use existing IT security and deployment methods
- Keep a low TCO

MEASURING YOUR CONTROL PROGRAMMING CAPABILITY

When we think about control system programming for PLCs, DCSs, and HMIs, we often tend to focus on the "control system" side, but it is important to remember that control system programming is programming. Most control system programming uses highly specialized IEC 61131-3 languages, such as ladder logic, sequential function charts, process flow diagrams, and function block diagrams. Despite the graphical nature of these languages, they are computer languages and the creation of the diagrams is a programming function. Usually the programming is not performed by programmers or software engineers but by chemical, industrial, electrical, or mechanical engineers or industrial technicians who have picked up programming though professional classes or after-graduation school work. However, just because the typical control system programmer is not a software engineer does not mean they cannot use software engineering methods to define, monitor, evaluate, and improve their work.

The quality of the final product, a control program, is largely determined by the quality of the development and maintenance processes used. The difference between organizations with well-defined and managed processes and those without is immense. Every year there are reports that on average only 30% of business IT projects finish on time and on budget. We regularly see major news reports about IT failures, such as the FBI write-off of their $170 million and 4-year development of a Virtual Case File system. Average errors per thousand lines of code and average lines of code per day, two common measures of programming productivity, vary by factors of over 100 between the best and the worst organizations. Organizations with well-defined and managed processes usually come within ±20% of estimated project time and effort, those without well-defined and managed processes often miss their times and costs by over 100%.

An important question for control system programmers is "how good are we?" Specifically, how well do we stack up against the average and against the best in class? This is a common question in business IT organizations and it is no less important for manufacturing IT. Fortunately, there is a generally recognized standard measure of software capability, the Software Engineering Institute's (SEI) Capability and Maturity Model (CMM). The first version was released in 1991 as a way for software companies to compare their practices and processes against the best in class. The latest incarnation is called the Capability and Maturity Model Integration (CMMI) (http://www.sei.cmu.edu/cmmi/), and it has been developed for software engineering, system engineering, integrated product and process development, and supplier sourcing. The ones most important for control system development is software engineering and system engineering (SE/SW) because of the typical hardware content in control systems.

The CMMI model has two types: staged and continuous. The continuous model focuses on organizational elements, such as project management, support, and engineering, and is generally appropriate for large organizations. The staged model focuses on the activities at each level and is generally appropriate for smaller organizations.

CMMI, described in Table 2, defines five levels of organizational maturity; 1—initial, 2—managed, 3—defined, 4—quantitatively managed, and 5—optimized and can be visualized as in Table 2. Each level defines activities

Table 2. CMMI Level

Level	Name	Description
1	Initial	Little or no rules for development
2	Managed	Process in place
3	Defined	Organization-wide process and methods
4	Quantitatively managed	Process measure performance and adaptable development
5	Optimized	Continual improvement process

and artifacts (documents, systems, and training) normally associated with execution at the level. Each level includes all of the tasks and artifacts of the lower level, and is an evolution of the lower level.

As an organization matures in its ability to develop, deliver, and maintain software, it should naturally increase its maturity level. However, you cannot jump levels, despite the attempt at many companies. Organizations at Level 1 often have spectacular successes and failures but cannot repeat their successes. Organizations at Level 5 are on time and on budget for even the hardest of tasks; they know their problem spaces and how to develop solutions using continual improvement methods.

Level 1 (initial)

Level 1 organizations have little or no rules for software development. They go from crisis to crisis, expending superhuman efforts during releases, and the success of a project is based on the skills of the developers. Released products may have one bug for every 1000 lines of code (or 1000 rungs of ladder logic). Projects are, on average, from 50% to 100% over time or budget.

Level 2 (managed)

In Level 2 organizations, software development is documented and repeatable. Level 2 organizations have processes in place for Requirements Management, Project Planning, Configuration Management, Software Quality Assurance, Subcontract Management, and Project Tracking. Level 2 organizations have a 25–50% reduction in code defects than Level 1 organizations. This reduces support costs and “emergency” fixes by over 50%. Fortunately, most groups developing control applications are at least Level 2 or should be. Projects are, on average, from 25% to 50% over time or budget. However, processes are project specific, and processes, management methods, and successes vary across the organization. The success of projects is based on the skills of the individual managers. Moving from Level 1 to Level 2 is not based on technology or development tools but on individual managers' commitment to improving quality and productivity.

Level 3 (defined)

Level 3 organizations use the same process and methods across the whole organization. They have defined and documented processes, with training to bring all management to a minimum level of competency. Processes typically include all Level 2 processes plus peer reviews, training programs, and formal cross-group coordination. Software development is repeatable across the organization and the best practices are used by all groups. Defect rates may be 25–50% lower than Level 2, further reducing maintenance efforts on the code and emergency fixes. Level 3 organizations follow their processes even under schedule pressures because management recognizes that it is the fastest way to finish. Projects are, on average, no more than 25% over time or budget. The success of projects is based on training and documentation of processes. Moving from Level 2 to Level 3 requires executive commitment to improving quality and productivity.

Level 4 (quantitatively managed)

Level 4 organizations extend their Level 3 capabilities to include management processes that measure process performance and adapt the development process for specific projects without losses of quality. Level 4 organizations have statistical control over the quality of processes and products, using SQC charts and other tools to quantitatively manage projects. Project problems are quickly recognized and corrective action taken. Error rates decrease but the real improvement is in the ability to meet schedules and budgets. Most projects finish on target and on time. The success of projects is based on the ability to objectively evaluate a project's performance and to have processes in place to take effective early action. Moving from Level 3 to Level 4 is based on a management commitment to finding problems before they get out of control.

Level 5 (optimized)

Level 5 organizations extend their Level 4 capabilities by adding continual improvement processes. These organizations use quantitative measures to proactively find the source of errors and correct the processes. Error rates may be as low as 1 defect per 1,000,000 lines of code. Six-sigma and other improvement processes are in place and are ingrained in the organizations culture. Projects are on target and on time. Only a small percentage of software teams ever reach Level 5, and those that do continually strive for further improvements.

Take it seriously

Control programming is a programming task, and organizations that take it seriously should know how they stack up against good and best-in-class organizations. Improving your capability is not solved using technology but is solved through management commitment to quality and executive involvement. See www.sei.scmu.edu for more information.

LET'S GET PERSONAL

The Software Engineering Institute's (SEI) Capability Model is good for helping an organization develop software on time and on budget; however, it does not provide much help for the individual control engineer or development team. Fortunately, there are two additional processes developed by SEI, one for individuals to improve their personal processes and the other for development teams to improve the team's development process. Individual and team measures of code quality and ability to meet schedules can vary widely. The SEI Personal Software Process (PSP) and Team Software Process (TSP) address individual and team quality.

Individual capability matters

While organizational capability is important, individual capability is critical to a quality product. It is not uncommon to see 10 to 1 differences in errors per thousand lines of code between the best and the worst software developers, and the best and the worst development teams. Fewer individual errors result in more predictable releases and higher productivity. The quality of the end product is determined by the process with the lowest quality; often these are individual or team processes.

PSP was developed by Watts Humphrey, the "2005 National Medal of Technology Laureate." Most professional engineers strive to improve the quality of their work and PSP shows software professionals how to approach their tasks in a professional manner. PSP brings a rigor to your personnel processes by defining how you must measure your own performance, track changes, analyze problems, learn from mistakes, and update your own personal processes based on lessons learned. TSP brings the same rigor to software teams.

Correct your bad habits

It is easy to fall into bad habits in software development. Everyone makes some mistakes during development, but few systematically learn from the mistakes to avoid repeating them. Yet learning from mistakes and improving the process is the basis for quality improvements.

Examples:
Consistently discovering the same errors during unit testing such as:

- un-initialized variables in C;
- reuse of latches in ladder;
- incomplete detailed design;
- misspelled variables; and
- incomplete understanding of the proper control algorithm.

PSP is the improvement process you can make to learn from these mistakes and modify your processes to reduce or eliminate them. This might involve earlier testing of smaller elements, external reviews, or earlier code walkthroughs. It is up to you to make changes to your development process to remove the source for errors or find them earlier.

Self-study for self-improvement

SEI provides a self-study PSP manual. This is not for the faint of heart; improving your own work is a long process but worth the journey. Improvements do not come easily, especially if you have been designing and implementing software your own way for a long time. The self-study works best when you are working on one or more large projects. It requires you to keep measures, through logs of all stages of a project (planning, design, code, compile, test, and maintenance). The logs allow you to establish a quality baseline, measure estimates of sizes, resources, and schedules, and practice defect and yield management.

The reward for this hard work is a better ability to meet your commitments by correctly estimating and planning your work. Engineers undergoing PSP training have reduced schedule estimating errors from an average of 350% (work took 3.5 times longer than expected) to less than 10%. Total defects per thousand lines of code were reduced by over 75%.

The team software process

The Team Software Process (TSP) provides similar process improvement tools to team managers, focusing on self-directed teams. It stresses team development of schedules and estimates. Software managers are considered coaches that motivate the team and maintain a focus on quality and excellence. TSP provides tools to help manages follow the process, gather data, analyze results, and make corrections. PSP and TSP together provide a basis for quality development.

Additional information is available at www.sei.cmu.edu/tsp in the paper "PSP: A Self-Improvement Process for Software Engineers."

Real-World Design

In the real world, IT solutions are usually effective for 3–5 years while manufacturing solutions usually need to last 10–20 years. This chapter presents some practical approaches to designing systems and support organizations that handle the long lifetimes of manufacturing IT systems.

The second section discusses the issues of designing for fallible systems. Even the best designed system can fail, but failure in manufacturing can lead to loss of products, physical damage of equipment, harm to humans, and even death. Manufacturing system failures often have serious consequences measured in millions of dollars, and this section presents some of the rules that can be applied when designing for a fallible world.

Methods to reduce the risk or the impact of failure are an important part of system design. The third section in this chapter defines some of the tools that one can use to understand the causes of failures and reduce their impact.

The last section in the chapter deals with the on-boarding process, which is the process that new employees or consultants are given to acquire the knowledge, tools, and access they need to be productive team members. This information may help to reduce the start-up time for new team members and make them more productive faster.

STABILITY AND CONSISTENCY IN THE CHANGING IT WORLD

A common problem for control engineers is how to design systems for long lifetimes, often 10–20 years, when using technologies that seem to change and become obsolete daily. The IT world has over 40 years of explosive growth and changes with quarterly upgrades in hardware, operating systems, and applications. It is difficult to buy the same IT hardware today that you bought last year, or even last month. Application software has one or two releases per year and loss of support after 3 years. Operating systems are a little better. The time from first release to out-of-support is about 5 years.

Unfortunately, all of these lifetimes are much shorter than the typical lifetime of a process control system. Yet, we now rely more on commercial off-the-shelf information technologies in control system solutions, and are asked to maintain and support IT-based systems well beyond their normal commercial lifetime when replacement hardware and software is not available and changes require revalidation and retesting.

Core change is difficult for IT

Effectively managing the life cycle of automation systems requires looking at all of the system's elements. In IT systems, there are core functions, which are difficult to change, and peripheral functions, which are easier to change. The typical IT system has core functions built around data structures and specialized applications. When these elements change, the effect can ripple through the entire system and changes are often visible to the user through new business processes, operating procedures, and application interfaces.

The OS, standard applications, and hardware are typically peripheral functions. The upgrade of a server OS (from Windows 2000 SP3 to Windows 2000 SP4 or even to Windows 2003) can be a minor event and changes in hardware are often completely invisible to the user. Even changes to standard packaged applications, such as word processors, spreadsheets, and browsers, usually have a minimal impact on users.

Peripheral change is difficult for manufacturing

In control applications, the situation is often reversed. In applications where SCADA, PLC, or PC-based control is used, a change in the hardware, OS, or packaged application usually has a major impact and may even cause the entire system to fail. Changes to these core functions require retesting and revalidation of the entire system.

However, changes to application data and special applications, such as PLC and SCADA programs, are often invisible to users. This is routinely performed and usually requires less retesting. This means the rules used in managing the IT systems' life cycle need to be reversed when managing process automation systems life cycle.

Hardware and OS changes are difficult in automation systems but easier in IT systems. Changes in application-specific programs and user-interface screens are common in automation systems but more difficult in IT systems.

Organizations that are set up to do life cycle management of process automation IT systems need to have a different support process, a different skill set loading, and probably a different organization than standard IT support organizations.

Looking for stability

Providing stability with continually changing IT elements requires identifying the systems that cannot change without significant impact and those that can change with minor impact.

Examples:

- Networks are peripheral if they are based on standard Ethernet technology. The basic network wiring, routers, and bridges can often be updated with minimal impact and limited testing.
- Database servers, files servers, and print servers can also often be updated with limited impact.

However, changes to packaged applications, such as HMI, historians, MES, and batch systems, can have significant impact and require major testing.

Systems that are core should be placed on a standard review cycle, such as once every two years, where each system is reviewed, life cycle support risks determined, and the decision to retain, upgrade, or replace is made.

Noncore systems can be reviewed on a shorter cycle or when upgrades are available, since upgrades are less severe.

Core elements will need a larger support organization that operates on a project basis, while peripheral systems will primarily have reactive support organizations. Creating stability and consistency in the IT world requires an attention to core life cycle support elements and continual review and evaluation of life cycle risk.

DESIGNING FOR A FALLIBLE WORLD

As control engineers, we tend to think that keeping the process running is our only job, but just because the process is operating does not mean that the plant is operating. We typically deal with IT elements needed to keep the control system running, such as firewalls and support activities, but that is not enough.

There is a larger issue that also needs to be addressed when using IT systems in manufacturing operations—a system architecture designed for IT failures.

The need for a robust system architecture was made obvious in 2003 when the MSBlast and LoveBug worms shut down IT systems around the world.

These events illustrated two types of manufacturing companies: those that continued production and those that had to shut down because of the worms or the IT response to the worms. Some companies were able to "raise the drawbridge" and continue production by disconnecting their operational systems from their business systems. Other companies could not separate their business and operational systems and had to shut down production. The different responses illustrated the difference between tightly coupled systems and loosely coupled systems and between global systems and local systems.

Many IT departments have an unspoken bias to tightly coupled global systems because they are generally easier to build and easier to maintain.

Example:

- Maintaining a single global instance of a document management system is easier that one instance per site.

In addition, if connections are needed to be made to other systems, it is easier to hardcode a tight synchronous connection than to design and implement a standard based asynchronous connection. When everything is working, global tightly coupled systems are fine, but when things go wrong, the effect can be global and catastrophic. The best approach to system design should be to emulate the well-known saying, "Expect the best but plan for the worst."

Separate critical systems

Companies that require absolutely reliable systems, such as banks and other financial institutions, will often set up separate servers, networks, and support organizations for the critical systems. These are separate from the normal business systems supporting HR, purchasing, and logistics.

This same approach is required for manufacturing companies when the IT systems are critical to maintaining plant operations. One of the first steps in this approach is to identify the systems critical to operations, and every system in use by operations needs to be examined.

Examples:

- A company is using a centralized configuration management server for controlling all changes to PLC code, then if that system becomes unavailable the maintenance group may be unable to make any emergency maintenance changes in PLC code during an IT outage.

- If a global license manager grants right-to-use for software packages and an IT outage occurs, then users may not be able to access displays, reports, or recipe management systems because they cannot obtain the right-to-use license.

Local systems for robust operations

When designing manufacturing systems that are robust, in the face of IT failure, a good approach is to use local systems instead of global systems. This eliminates outages because of WAN failures, which were a major cause of the plant shutdowns during the MSBlaster attack.

Multiple local systems are more expensive to maintain, but they are much more robust in the face of failures. Another good approach for designing robust systems is to use system interfaces that are asynchronous and buffered. This approach allows for temporary loss of communications or system failures without causing cascaded system failures. Interfaces based on messaging systems are especially robust in the face of network and application failures. These interfaces should be the default choice for systems that do not need real-time synchronous communication.

Fortunately, not all solutions to robust operations in the face of IT outages need to be technology oriented. Phone lists and faxes can be used to collect critical decision information, and paper backup systems can be used to record critical information. However, the worst situation is to design an operational system based on the assumption that there will be no failures in network infrastructure or in IT applications.

FOCUS ON AVOIDANCE, NOT ON THE RISK

When riding my bicycle wearing my bright yellow jersey and flashing lights, most drivers will make sure that they pass safely, but some will focus on the yellow jersey and flashing lights and slowly drift into the bike lane. These people are focusing on the risk, not focusing on the actions needed to avoid the risk, and their lack of actions actually increase risk. In motorcycle riding, there is a similar situation. If you focus on the tree you might to run into or the parked cars you might hit, instead of focusing on turning and stopping, you will invariably hit the tree or parked cars. The rule is to focus on the actions to avoid risks, not just focus on the risks.

Don't be paralyzed by risks

The advice to focus on the actions of avoidance and not the risks also applies to manufacturing and automation systems. There are a multitude of risks to manufacturing computer systems including: loss of electrical power, poor quality electricity, hardware failures, software failures, viruses, hackers, and environmental hazards (flood, fire, wind, earthquakes). While all of these risks can be real, it is important to remember that multiple risks can be addressed if you focus on the actions needed to avoid risks. Some IT organizations will spend a large amount of time and effort documenting all possible risks, generating reams of paper that list all of the possible ways the system could become unavailable. Yet these same organizations do not apply a systematic approach to risk identification that combines risks into groups that are mitigated through a common avoidance strategy. A focus on risks can paralyze an organization, preventing even the simplest actions to avoid risk because the actions don't avoid all risks.

Tools for risk analysis

There are three tools that can help categorize risks and provide insight into single actions that can mitigate multiple risks. These are Ishikawa Diagrams (also known as fishbone diagrams), Fault Tree Analysis (FTA), and Failure Mode and Effect Analysis (FMEA). Ishikawa Diagrams were named after Professor Kaoru Ishikawa, a pioneer of quality management who devised this representation in the 1960s (www.wikipedia.org/wiki/Ishikawa_diagram). The diagram in Figure 2 resembles a fish skeleton, when complete with major branches for areas of risks and end points as unique risks. It is commonly called a Fishbone Diagram.

Usually a team will generate the diagram and then rank the probability, severity, and ability to detect the failure. Failure Mode Effect Analysis (FEMA) and Fault Tree Analysis (FTA) are tools to help in risk analysis

An Ishikawa diagram lists areas of the system where risks can occur, namely personnel, environment, software, and so on. A Fault Tree Analysis is used to determine the events or conditions that will lead to a fault or failure (www.wikipedia.org/wiki/Fault_tree_analysis). It is effective at documenting parallel events or sequential events that can lead to a fault.

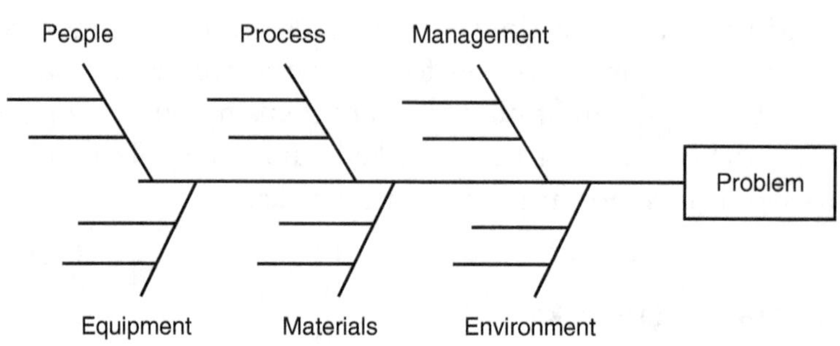

Figure 2. Fishbone diagram.

FEMA is a rigorous approach for identifying all possible failures (www.wikipedia.org/wiki/Failure_mode_and_effects_analysis). It is used to document current knowledge and all approaches to reduce or prevent failures.

Ishikawa diagrams, FTA, and FEMA are all tools to help understand the risks, and should also be used to understand the actions needed to eliminate risks. In particular, manufacturing IT systems face risks due to external threats, electrical power loss, hardware failures, physical environment, and software failures.

Simple actions to minimize risks

It may be impossible to eliminate all of the external threats to your manufacturing systems, but simple actions can be used to minimize the risks. Only allow outside access through firewalls and tightly constrained proxy servers with a Demilitarized Zone (DMZ). A DMZ isolates critical systems from other corporate systems and Internet access. These actions do not mean you can avoid external risks, but they will reduce your exposure to these risks.

Electrical power risks can be minimized through use of uninterruptable power supply (UPS) systems and industrial-grade surge protectors. UPS systems should be connected to server systems and workstations through USB ports to allow orderly shutdown and preservation of critical data. UPS system tests must be added to normal maintenance tests. UPS batteries last only a few years, so they must be regularly tested and replaced. The worst UPS system to have is one that doesn't work when needed.

Hardware failure risk can be reduced by focusing on hot, warm, or cold standby systems. The decision of which standby method to use should be based on the criticality of control and acceptable business downtime. Many production facilities will experience hundreds of thousands, if not millions, of dollars of losses for each day down. A small fraction of that amount spent on avoidance can reduce the risk.

Hot standby systems are the most expensive, but they have switchover times of milliseconds to seconds. Warm standby systems are typically implemented using ready-to-run copies of the primary systems and have switchover times of minutes. Cold standby systems have the same hardware configuration as the primary but have no software. When a primary system fails, the standby is loaded with a ghost image of the primary system. Usually one standby can cover multiple primary systems. Cold standby switchover times are usually measured in hours.

Environmental hazards can be avoided by having an alternate backup site. The alternate site should be physically separated and isolated from the primary site. It should be in a separate building, on higher floors to avoid floods or lower floors to avoid wind, with separate power lines and network communications. Alternate sites should have the capacity to run for extended periods of time, even if at a reduced performance.

Software risks can be reduced by a philosophy of "test-first, test-always" for patches and updates. A common cause of software failure is from patches and updates that change undocumented features that were used by application software. Cold standby systems are often used to test patches and updates, reducing the cost of software risk avoidance. The best failure avoidance strategy is to "always, always, always test before you install in a production environment."

It is important to be aware of the risks to manufacturing and automation systems, but it is more important to focus on the actions required to avoid the risks. By focusing on avoidance actions your systems will become more stable and you will have fewer chances of stopping production due to an avoidable risk.

STARTING OFF ON THE RIGHT FOOT

With so many automation and MES projects being implemented by consultants and contractors, it is important to have good processes in place to make them productive as quickly as possible. In many projects, the on-boarding time, which is the time to set up all of the proper hardware, accounts and privileges, and ramp-up time, which is the time it takes for someone to start doing useful work on the project, can be a month or more. This month is nonproductive time and can be a substantial part of the project cost.

Getting on-board

Most IT departments have on-boarding processes for getting PCs, user accounts, and standard tools. These processes work well for on-boarding administrative and clerical personnel, but workers in automation projects often need additional specialized hardware, software tools, and IT administration rights. Automation contractors also need to know your automation project processes and their roles in the processes. It is important that this information is readily available in order to reduce the on-boarding and ramp-up time. It is even more important to have easily available information if the contractors are off-shore. Because of time zone differences and potential language problems with off-shore contractors, simple misunderstandings of project process rules can take days to be resolved, slowing down ramp-up and negatively impacting the project's schedule.

On-boarding for automation projects should augment the IT on-boarding process by adding other necessary components such as:

- PLC or DCS programming tools,
- fieldbus network cards in PCs or laptops,
- configuration management accounts for code check-in and check-out,
- defect tracking accounts,
- test system accounts,
- documentation system accounts,
- project management tools,
- physical access rights, through key cards, PIN numbers, or biometric authentication.

IT organizations often have separate on-boarding processes for their own staff, which usually includes additional administrative rights and software tools. Manufacturing managers should work with their IT organization to add standard on-boarding processes for automation and MES contractors. Ensuring that contractors have the right tools will eliminate problems later in the project.

Learn the processes

Once your automation resources are fully on-boarded, they will need to understand your development processes. Fortunately, it does not take sophisticated tools to provide this information. Two easily implemented methods are through collaboration services website or a Wiki system that provides on-boarding and ramp-up information.

The website or Wiki should start with a list of general questions that are always asked by on-boarding personnel. These would include:

- "What do I do first?"
- "How do I see the tasks that have been assigned to me?"
- "How do I see the meeting schedule?"
- "How do I submit a problem report on (O/S, project code, tools)?"
- "What is the project schedule?"
- "What is the holiday schedule" (this is very important and often overlooked when using off-shore consultants)?
- "What is the meeting schedule?"
- "What is the project organization chart?"
- "Who are the vendor contacts?"
- and the most important "What are all of those three-letter acronyms?"

The answer to each question should include a graphical representation of the process, an identification of who to contact for questions on the process, links to the appropriate internal website for additional information, and links to internal tools used to initiate a process or take an action. The advantage of using a Wiki is that changes can be quickly accomplished without calling on IT resources. A Wiki can also be quickly extended for project-specific information and processes.

Don't underestimate the ramp-up time

The on-boarding and ramp-up time for contractors are usually built into project schedules, but are often underestimated. Spending time before any project starts to work with the IT organization to define the tools and processes, your contractors need to save man-months of unproductive time, time that your project is paying for.

Tools and Languages

This chapter covers IT tools that are commonly used in medium- to large-size companies. These tools are usually so pervasive that they are part of every IT project, and this increasingly includes manufacturing IT projects. Therefore, understanding IT tools is increasingly important for manufacturing engineers.

The first section in the chapter discusses Enterprise Service Buses, which are a set of servers and web services that pass messages between programs. Enterprise Service Buses provide a centralized mechanism to handle the complex communication paths normally involved in large system integration projects. You will probably encounter an Enterprise Service Bus in manufacturing to business integration projects.

Manufacturing Integration Buses are the manufacturing equivalent of Enterprise Service Buses and are defined in the second section. These buses use a set of standards and tools that are unique to the manufacturing environment. A Manufacturing Integration Buses can complement a corporate Enterprise Service Bus, but its value must often be explained to corporate IT departments.

The next concept described in this chapter is Master Data Management. Master Data Management defines a process used to manage all of the computer-based information required to operate a business, such as material codes and product codes. Manufacturing systems often need to use, manipulate, and understand these data, so it is important to know how Master Data Management fits into your processes.

There are many practical considerations involved in developing Manufacturing IT systems, and one of the most important is selection of the correct computer language to use. Every language has specific strengths and weaknesses and it is vital to pick the right language for the task to be performed. In many applications, multiple languages may be justified, each for a different part of the system, and simple rules described in this chapter will help in your language selection process.

The last concept is cloud computing. This is a common model in IT where computing resources are outsourced and reside in large, remotely accessed server farms. Cloud systems provide great economies of scale, and can be effectively be used in manufacturing systems.

RIDING THE ENTERPRISE SERVICE BUS

The typical IT environment is a federation of systems. The term "federation" in the IT world is applied to collections of applications from multiple vendors that work together to support business processes. A federation may include separate applications for material management, order processing, supply chain management, customer relations, and production scheduling. Even when a company has selected a primary ERP vendor, there is often a federation of legacy systems supporting unique business processes. Federated systems are expensive and integration efforts are often a major portion of IT budgets.

An increasingly common method to reduce integration costs is an Enterprise Service Bus (ESB), sometimes called an Enterprise Integration Bus (EIB). These are not electronic buses in the sense of an electrical backplane bus. Instead, they are specialized applications that run on redundant servers and act as concentrators and distributors of data. Manufacturing systems that must exchange data with business systems will probably need to connect to the company's ESB.

ESB is a concept not a standard

ESB are an architectural concept that includes open standards, message-based communications, message routing capabilities, and service discovery mechanisms. There is no single definition of an ESB product, but a working rule is that it is a system that provides a single source of shared information, a single location for discovering application services, and a single destination for using services.

Several vendors provide ESB, but a few manufacturing companies have also built their own ESB systems based on open standards and focused on their unique integration problems. Once a company has selected an ESB system, the IT department will attempt to have all applications that exchange data (including manufacturing applications) use the ESB instead of implementing point-to-point connections. Unfortunately, there is little interoperability among different ESB systems, so each application interface must be customized for the chosen ESB.

There are five main elements of ESBs that are important in connecting applications to an ESB. All of the elements are based on XML technologies and most ESBs are based on web services.

1. A **transport** element to move messages.
2. A **discovery** element to discover where to send the messages.
3. A **transform** element to convert messages to and from application specific formats.
4. A **protocol** element to handle transactions associated with messages.
5. A **payload** management element to handle the XML messages.

Transport

The transport element handles transporting XML messages from one application to another through the common server. This eliminates point-to-point interfaces and provides a centralized mechanism to manage and view interapplication communication.

HTTP messages and Java Message Services (JMS) are common open-source implementations transport layer implementations. The transport element provides guaranteed delivery of messages, queuing messages in databases on the ESB servers if the receiver or a downstream ESB server is not available to receive the message.

Guaranteed message delivery is a key requirement of any ESB.

Discovery

The discovery element allows applications to discover the services and data provided by the ESB. This is typically handled by UDDI services (uddi.xml.org) in the IT environment. There are many different discovery methods, but a good ESB will provide a method to discover where to send a message based on the service that is needed.

In an ESB environment, applications do not directly communicate to other applications, but they discover who to send the message to at run time on the basis of what application is providing the service they need. This method of run-time discovery means that applications can be removed and replaced without breaking hardcoded links to other system.

Run-time discovery is a key feature of any ESB.

Transform

The transform element provides methods that convert data from the sender's format to the receiver's format through a set of application-specific transform rules. This can be performed using some form of XML transformation, such as XSLT scripts (www.wikipedia.org/wiki/Xslt).

Transformation can be a complex task. So many companies use an intermediate representation for messages, known as a canonical form. The ESB transform services convert messages received into the canonical form. The canonical messages are sent through the ESB to the receiver, where the ESB transform services convert them into the receiver's format.

In the manufacturing area, the ISA 95 Enterprise/Control Integration standard with the associated MESA B2MML schemas (www.mesa.org) are the most commonly used standards. Many ESBs will provide a built-in transform layer that can change an application specific message to a correctly formatted standard message.

Protocol

The protocol element checks the formal definition of allowable message transactions and rejects messages that do not match the definition rules. For example, a GET message will be followed by a SHOW response to the original sender. The protocol elements of an ESB will maintain information about the transaction messages to line-up senders and receivers.

Many protocol and transaction rules sets are defined in XSD formats based on standards such as OAGIS (www.oagi.org), RosettaNet (www.rosettanet.org), and the ISA 95 B2MML transaction definitions.

Payload

The payload in an ESB is the data that makes up the body of the message. ESB messages are XML documents, which can be very long. Many ESBs will provide methods to compress before sending and storing, and then uncompress the messages before delivery to the receiver. ESB messages in the manufacturing area can be megabytes in length, especially messages that deal with materials and their properties. It is important to estimate the message rate and size you expect to see when discussing your integration needs with your IT group. They may need to increase the size of the ESB servers and server storage to handle manufacturing application integration. The payload management element may also provide multiple methods to send the message to the ESB and receive a message, such as through a file as an in-memory Document Object Model (DOM) object.

Get on the selection bus

If your company has not yet selected an ESB, then make sure that the Manufacturing IT group is involved in any selection decision. If your manufacturing applications are not compatible with the ESB selected, then custom interfaces will need to be developed by you or your vendor, increasing your cost to ride the Enterprise System Bus.

THE MIB IS HERE TO HELP

We have just described Enterprise Service Buses (ESBs) architectures and how they are being used for business system integration. ESB architectures can also be valuable tools within the manufacturing control environment. A Manufacturing Integration Bus (MIB) is an ESB that is designed for the exchange of information between manufacturing applications. These applications include SCADA, MES, LIMS, data historians, Asset Management systems, warehouse management, and CAPA systems.

Standards for the bus

There are several existing and emerging manufacturing standards that enable MIBs. The existing standards are ISA 95 Enterprise/Control System Integration, MESA Business to Manufacturing Markup Language (B2MML), and OPC-UA. The ISA 95 Integration standard defines the standard terminology for typical information exchanges about material, personal, equipment, physical assets, production schedules, and production responses. The MESA B2MML is an XML implementation of the ISA 95 Integration standard and defines the exact format for representing exchanged information.

OPC-UA (Unified Architecture) (www.opcfoundation.org) is a vendor-independent message transport mechanism. It provides a cohesive, secure, and reliable cross-platform framework for access to real-time and historical data and events. OPC-UA is designed for high reliability and redundancy, with configurable timeouts, error detection, and recovery features to deal with any error or fault conditions (such as the loss of network communication).

Also common on MIBs are two standards, the ISA 95.05 Business to Manufacturing Transaction standard and the MESA B2MML release, that support the transactions.

Push/pull and publish/subscribe protocols

The ISA 95 Transaction standard defines message transactions that can be used for MIB web services. The services have three methods for information exchange:

- Pull methods where applications ask owners for data.
- Push methods where applications push new and changed data to data owners.
- Publish/subscribe methods that allow multiple applications to keep information synchronized.

MESA B2MML version 4.0 and later are the XML implementation of the ISA 95 Transaction standard and are also consistent with the business industry OAGIS 9.0 specification (Open Applications Group Interface Standard) and the UN/CEFACT Core Components (United Nations Center for Trade Facilities and Electronic Business).

The MESA schemas define the formal structure for MIB web services data. OPC-UA defines a platform-independent transport mechanism for XML messages and defines vendor neutral methods for interfacing to the MIB. Applications would use OPC-UA to exchange messages between clients and servers over various types of networks. Servers define the object models that can contain server-defined data types. OPC-UA also defines standard services for servers. Two transport mechanisms, TCP and SOAP using HTTP, are defined in the initial OPC-UA documentation.

MIB and OPC-UA

The MIB is both an OPC-UA client and a server. It is a client for owners of data and it is a server of that data to other clients. Manufacturing and business applications that are providers of data would register the data classes with the MIB that they provide (such as personnel, material, production schedules, etc.) and what services are provided for the data.

MIB in the DMZ

An MIB also solves two other major Manufacturing IT problems: security and ERP communication failures. An MIB server placed within a manufacturing zone DMZ eliminates direct business system to control system communication and closes many associated security holes.

MIB handling of communication failures

When the MIB also includes data caching, then it handles failures in ERP communication. If the ERP system or other business systems are unavailable, then critical manufacturing systems can operate from cached data. The MIB would subscribe to all registered business data and would cache published changes. When an application pulls data, the MIB would respond with data from the local cache.

MIB simplifies integration

Manufacturing Integration Buses provide valuable tools for manufacturing IT. They decouple manufacturing applications, simplify business to manufacturing integration, improve manufacturing system security, and handle ERP communication failures.

MIBs can be constructed using open source or commercially available tools and industry standard elements, such as XSLT, JMS (Java Message System), SOAP, HTTP, B2MML, and OPC-UA. Any company that is updating its manufacturing IT environment should consider an MIB as the core backbone for manufacturing application integration.

MANUFACTURING IN THE CLOUDS

The term "cloud computing" is frequently used in IT departments, and "in the cloud" has even become a common definition of where software or data exist. The image of a cloud has been widely used in network architecture diagrams to represent the Internet. Cloud computing simply means using applications and services that are accessed using the Internet. Cloud computing is essentially a method for outsourcing a complete IT service where the service user, the IT department, does not really know or care where the service is hosted. Examples of IT services that are available "in the cloud" are sales force automation, customer relationship management, payroll, taxes, e-commerce, accounting, logistics, shipping management, receiving management, invoicing, and data backups.

Cloud computing service providers use computer clusters and large server farms to meet expected demand. Server farms may have thousands of physical servers and tens of thousands of virtual servers, with built-in power and network access redundancy. When a cloud computing service provider has a large number of users, it can offer inexpensive services because of economies of scale. There are over 50 different cloud computing service providers, including powerhouses such as Amazon, Google, Microsoft, and IBM, providing a lot of opportunities for economies of scale.

Cloud computing users also gain some significant economic advantages. Cloud computing users have no capital expenses, they have reduced service costs because of a simplified IT infrastructure, and there is often a reduction in cost of large client applications, because of reduced maintenance fees. Cloud computing users also usually have the ability to rapidly scale up and scale down the number of servers and amount of service that they need. For seasonable companies, this means that can scale up during their season, such as the Christmas holidays, and scale down after their season. Cloud computing users do not have to buy and operate systems scaled to their worst-case use scenarios. The primary disadvantages of cloud computing systems are the risks associated with user access to the Internet and the financial stability of the service provider.

Cloud computing can take many different forms:

- Applications in the cloud: In this model the application software, such as a database manager, office tool, or payroll application, resides on the cloud servers. The data reside on the user company's servers so that it is easier to protect the data. This model is often used to manage a large number of end users and simplify maintenance and updates to the applications.
- Data in the cloud: In this model the data used by applications reside on the cloud servers while the applications run on local systems, such as office tools on the end user workstations and laptops. This model is often used to provide redundancy and backup of databases.
- Server in the cloud: In this model the applications and data reside on servers in the cloud. End-user workstations simply act as display terminals, with all computing occurring on the cloud servers. This is the maximum outsourced model with the minimal cost for end user workstations.

As cloud computing has grown, IT managers and CTOs often ask if manufacturing can operate in the cloud, and manufacturing organizations should have an answer. The typical cloud computing application adds a lot of risk in a manufacturing environment. The service is accessible through an uncontrolled network (the Internet), and there is little control of external problems, such as bankruptcy, flood, fire, earthquakes, and weather disasters. While cloud computing vendors have backups and redundancy, the user company's own internet connection may be the weak link in accessing the cloud service. These problems mean that cloud computing should not be used for mission critical applications. However, manufacturing organizations can use the concepts of cloud computing in mission critical applications in a local mode.

If you remember your science from fifth grade you can use cloud types to describe different types of cloud computing.

- The highest clouds are cirrus clouds and the cirrus cloud model can be visualized as hosted services available from a cloud computing vendor through the Internet. The cirrus cloud computing model is suitable for non-mission-critical applications, such as document management, analysis applications, and system backup. These services can often be offline for short periods of time and do not require 24 × 7 accessibility.
- Stratus clouds are intermediate-level clouds and can represent the cloud computing model used within your company's business network. In this model your company owns the server farm, but provides the same level and type of services that an external vendor would provide. This model provides the economies-of-scale advantages while retaining control of the applications and

data. Applications that are not business critical can follow a stratus cloud computing model. These would be applications hosted through the company's intranet in a single server cluster. This is often the model used for corporate e-mail, controlled document management, document workflow applications, and knowledge management applications.

- Cumulus clouds are low-lying clouds and can represent the cloud computing model used within your company's manufacturing operations network. In this model you would have a division- or site-level server farm that provides a similar level and type of services that an external vendor would provide. Because the applications are focused on manufacturing, the backup and support services can be designed using production system requirements.

Applications that are manufacturing mission critical and that normally run in a company's manufacturing operations network can follow the cumulus cloud model. In this model all of the manufacturing operations networks within a site would be combined into a single network. This is a network that is firewalled and access protected from the business network with no direct access to the Internet. Common applications would then be hosted from the manufacturing cloud. Examples of these applications include MES, SCADA, HMI, LIMS, and Data Historian systems. There are significant advantages of scale if you can provide site-wide services from a single server farm. However, not all applications can operate in a remote server farm due to network restrictions or connectivity to plant floor equipment, so check with your vendors to see which applications can be consolidated into a single manufacturing server farm. You can answer questions about manufacturing computing in the cloud by creating your own site-level computing cloud for mission-critical manufacturing applications.

DO YOU SPEAK MY LANGUAGE?

It is a lucky individual who already has a computer science degree, or at least some programming classes under one's belt, when a programming task arises. Often, the task falls to those whose education is in other fields, such as physics, chemistry, biology, chemical engineering, mechanical engineering, or electrical engineering, and who have the need and the knowledge of the task at hand to program.

The first question these neophyte programmers ask is what language they should learn and use: thousands exist out there. Without a good grasp of programming, they will be hard pressed to determine the "right" answer to this question.

Set a language policy

The ideal solution to this problem is to have a policy in place about language choice. The policy will either suggest or require the appropriate languages to be used. There may be more than one language specified in the policy, because different languages may be the best choice for different problems. A well-thought-out policy will take into account various factors; the problem to be solved, the available hardware, available vendor-supported languages, industry standards, and legacy code already in use.

Creativity can make the most unsuited language usable for any task; after all, engineers are by nature creative individuals. However, by making well-thought-out judgments in language selections in the policy, the amount of time spent programming solutions can be greatly reduced.

- If the application contains a significant user interface and requires a small footprint on the user system, then JavaScript and the AJAX model (Asynchronous JavaScript and XML) for the client is currently the best solution.
- If the application contains a significant user interface and installation on the user system is possible, then Visual Basic or some form of Basic may be the best language.
- If the application has a database and web server side, then a scripting language such as Perl, PHP, Python, or Ruby is a good solution.
- If the application is heavily communication- and database-oriented then a strongly type-checked object-oriented language such as C++, C#, or Java are good solutions.
- If the application is real time or near real time then a language such as C++ is a good solution because of the ability to control execution cycles and enforce strong type checking.
- If the application is hard real time, requiring millisecond control of cycles and loops, then the IEC 61131-3 programming languages are the current best solution.

In general, compiled languages like C++ tend to run faster than noncompiled languages such as Perl or PHP, so if the speed of the application is critical, then compiled languages should be considered. Although Java and Visual Basic are often considered interpreted languages, they are actually compiled into machine-independent intermediate instructions that execute nearly as fast as fully compiled languages.

Complex is slow, simple is fast

Also consider the rule of thumb that the time to develop and test a program scales (at best) to the length of the program squared ($Time = ProgramLength^2$).

Compiled languages usually contain much more code to perform the same task than noncompiled languages. For programs that are run infrequently and are not under time constraints, the execution time saved by compiled than by noncompiled languages may not be worth the extra time spent in development. Additionally, sufficiently powerful hardware can be used to decrease program run time. In the tradeoff of faster hardware versus software development time, faster hardware is usually cheaper in the manufacturing environment.

Strongly typed languages and standards for languages

When picking a language, if possible, always choose the strongest typed language. This may seem counterintuitive, because strong type checking requires that every variable must be declared and this adds to the program length. Strongly typed languages or languages that provide an option for forced type checking (such as Visual Basic) will expose coding errors much earlier in the project than weakly typed, or nontyped languages. Many problems related to the use of an incorrect variable type can only be discovered at run time in a nontyped language. The extra time spent in coding to a strongly typed language will be made up in the testing phase of the project.

Ideally, the language you choose will also be an industry standard or a national/international standard. This allows for easier implementation of available manufacturing solutions, as well as the ability to use solutions already developed by others. If you use a vendor proprietary language, then support costs will often exceed any short-term development advantage. The average lifetime of manufacturing systems is measured in decades. Think about the maintenance effort required 10 or 15 years from now, when the vendor may be out of business and all "experts" in the language have retired. Industry standard languages will significantly increase your chance of finding someone to maintain the system in the future.

A policy about programming should also cover documentation of designs (which are often driven by the capability of the programming language), documentation of the code (including source code commenting), and policies on how to store useful code pieces such as subroutines or objects. Again, it is important to remember that those doing the majority of the programming are probably not well versed in these practices, so the policy should describe the advantages, such as the savings in man-hours and reduction in maintenance headaches, by following the policy.

Special policy for MULEs

Consideration should also be made for legacy code. Legacy code is inherited from former projects or employees who remain vital to operations. Ideally, you want to be able to maintain this code, or if you're putting into place a policy about programming languages, make arrangements for the code to be updated to a compliant language. If you cannot maintain or update these legacy systems, then they will be a MULES (Mature Undocumented Legacy Execution System), and will exhibit all of the stubbornness of a mule as you try to move them forward. Keep your software from becoming MULES by using the right programming language at the start.

The right language is worth the price

Picking the right language for a manufacturing IT project may seem like a minor exercise, but it can have a major impact in your schedule and the maintainability of the code. If you choose wisely, you will reap the rewards in future years. Your projects will be self-documented (through code comment standards), you will have a supply of people who can read and maintain the project, and you will be using a language that is appropriate for the job.

MANUFACTURING INTELLIGENCE FOR SMART MANUFACTURING DECISIONS

All companies are trying to make better business decisions and they are using Business Intelligence tools to help them. Business Intelligence (BI) is the term given to a set of tools and processes used to make business decisions based on facts instead of guesses, intuition, and often irrelevant experience. Few companies can afford to guess, so decisions based on sound data and provable relationships are critical for business survival.

Business intelligence for business decisions

BI was coined in the late 1990s by the IT community and has come to describe the tools used for the collection of business data, integration of the data across multiple dimensions, analysis of the data, and presentation of the results. BI tools are the preferred method for extracting information and correlations from massive databases of customer and sales data. BI analysis typically focuses on data about sales, customers, suppliers, shipments, purchase orders, invoices, and deliveries.

Manufacturing intelligence for manufacturing decisions

The manufacturing equivalent of Business Intelligence is Manufacturing Intelligence (MI). Most manufacturing companies have engineers who have considerable knowledge and experience in extracting information and correlations from production data using data historians and associated analysis tools. MI expands the concepts of traditional studies by looking across multiple dimensions of production information to discover correlations not seen in normal engineering analysis.

MI focuses on products, inventory, WIP, production schedules, production performance, yields, waste, and costs. MI usually does not need the high-resolution data collected in data historians. The typical MI time frame may be a production step, duration of a production run, or length of a batch operation. MI, like BI, is based on multidimensional data and advanced query and reporting systems. Multidimensional data are maintained in "facts" and "dimension" tables.

Just the facts

Facts are measurable pieces of information, such as a production yield or product quality measure. Each fact is associated with multiple dimensions, such as time, batch number, production line, shift, product supplier, batch step, and laboratory ID. An MI database can be huge with billions of facts and dozens of dimensions per fact.

The analysis capability required for MI is the same as the capability required for BI, but BI tools can be expensive and hard to justify in today's economic environment. Fortunately, there are low-priced alternatives. Inexpensive servers built on recycled MS Windows machines running the Ubuntu (A Linux system) can provide an inexpensive platform for open-source BI tools. Any Internet search will turn up multiple open-source BI tool sets, but there are a few tools that stand out. The Business Intelligence and Reporting Tools (BIRT) project is an Eclipse-based set of tools available at www.eclipse.org/birt/phoenix. BIRT provides Eclipse plug-ins and Java/J2EE code examples for BI analysis. Jaspersoft (www.jaspersoft.com) and Pentaho (www.pentaho.com) also offer free open-source BI tools. Any of these tools provide a good starting point for an inexpensive MI system.

Collect real data

The typical MI system would have server-side software that collects facts and dimension data from data historians, MES, and LIMS databases on a daily basis. MI is not designed for real-time decision making, so collecting data in a batch mode can be used to simplify data collection and storage. Perl, PHP, or Python programs would be used to quickly pull data using SQL queries and place the data into an MI fact-and-dimension database. The open-source reporting tools would then be used to run analysis and develop correlations for intelligent manufacturing decisions.

Data to knowledge

Implementing an MI program to augment your Six Sigma and other improvement projects does not have to be a high-cost project. Medium performance Linux-based servers running open-source software are the tools to help turn your manufacturing data into manufacturing knowledge.

Practical Programming

This chapter is concerned with MSI elements of practical programming, programming environments, programming project strategies, and programming diagnostics. Working in a collaborative environment can be challenging, but a Continual Build and Test environment, described in this chapter, can simplify development and improve release decisions. A Continual Build and Test environment provides a method to always have "ready to release" systems.

Large projects often have a difficult time "moving the ball forward," because of changing user requirements and resulting changes to existing code. A method call SCRUM using the Agile Software Development methodology is described here. SCRUM and Agile define an innovative new way to approach software projects when requirements are evolving.

Every engineer needs a toolbox to support practical problems, and engineers working in Manufacturing IT are no different. The next section in this chapter defines some of the practical software tools that should be in any manufacturing IT professional's toolbox.

Implementing large systems usually involved multiple servers and networks which must be separately specified, purchased, installed and maintained. The IT concept of virtual systems provides a good solution for reducing the effort required to install, maintain, and upgrade manufacturing IT systems. Virtualization also provides other advantages for security, patching, and redundancy. The last sections in this chapter define the methods and advantages for using virtualization in a manufacturing environment.

PLAYING IN YOUR SANDBOX

Control system development is usually performed by a small group. However, techniques that are commonly used by large IT development groups can also be used in small manufacturing IT groups. If you are using a waterfall development method (www.wikipedia.org/wiki/Waterfall_model), or the SCRUM (www.scrumalliance.org) and Agile (www.agilealliance.org) methods, there are well-accepted techniques for handling unit testing, system testing, Final Acceptance Tests (FAT) and Site Acceptance Tests (SAT). This can be important because there is often a lot of management pressure to move to an SAT or FAT before you have finished system testing. Unlike other engineering disciplines, in software engineering, it is difficult to statically evaluate software and see if it is ready for use. It has to be running to see what actually works, so management has no way to see if the project is "good enough."

When is "Good enough," good enough?

Deployment in a real system is often the only way to see if it is good enough for use in a real system, so this is a Catch-22 problem. "Good enough" means that enough of the basic functionality is available to be useful, even if all of the specified functionality is not yet working. Because of this pressure to deploy and evaluate, sometimes software is sent to a site or placed into use that has not gone through system or integration testing.

It is also very common in projects to "reevaluate priorities" so that software can be deployed without all of the initially defined features. This also often means the delayed features are never added because of the additional costs of repeating system tests, FATs and SATs. It also means that hidden dependencies between implemented and nonimplemented parts are not discovered until run time.

Continual build and test environment

One way to avoid this problem is through a Continual Build and Test (CBT) cycle and the use of multiple build directories. Using a CBT cycle means that you will always have a "ready-to-ship" system; however, it is a system that does not have complete functionality.

CBT starts by using an automated build and test system. The tests are Build Tests (BTs), which include the functions of integration testing, system testing, and FATs. Automation of the CBT is through "make" files, "Ant" files, or command scripts are used to automatically build the software and then run a defined set of tests.

The best method is to run the CBT cycle every night with the automated BTs used to validate the nightly build. When a new function or feature is added, then tests must also be developed. When a feature set is released to the nightly build directory, then BTs must be part of the release.

Continual builds may seem to be incompatible with a waterfall method of development, but the CBT approach should be used during the coding and testing phases of the project. The CBT approach provides continual system or integration test functionality, because the entire system is integrated and tested every night. Every build adds features to the system, bringing it closer to the "good enough" state for release.

Don't break the build

In the CBT approach, there is a lot of pressure to not "break-the-build." All developers have the responsibility for running unit tests, ensuring that their code builds, and developing the BTs, and should have a good expectation that their code will pass the BT. Developers perform these tests in their own local "sandbox" directory.

Everything that a developer works on is in their own sandbox, while the files they are not modifying come from the latest nightly build directory. This does make the "make" files more complex because the build utility must check the local sandbox for files and then check the build directory for the unmodified files. When a developer has finished coding and unit testing, he or she must check in the code and the updated BTs to test the code in the system. It is also a good practice to add documentation or notes that indicate what features were added or modified to help debugging if problems occur in the build or BT.

All developers have responsibility for checking their part of the build and addressing any problems that came up during the FAT run in the morning after a CBT cycle. Once the daily build has been successfully built and the BTs successfully run, the build directory should be moved to a "user test" directory.

The same CBT cycle should be run on the directory and the resulting system made available for user tests. The user test directory is always a "ready-to-deliver" system. At predefined intervals, or when the required level of functionality is reached, the user test directory files can be copied to a "Final Build" directory, and the CBT cycle is run again.

The final build is now ready for FATs, SATs, and distribution. Using the CBT method, the question never comes up if testing should be cut short. Instead the question is: "Are the functions and features already in place and tested good enough for release?"

CBT means always being ready to ship

The CBT approach, as depicted in Figure 3, is one of the most successful methods used in large projects, but it should also be used on small projects to

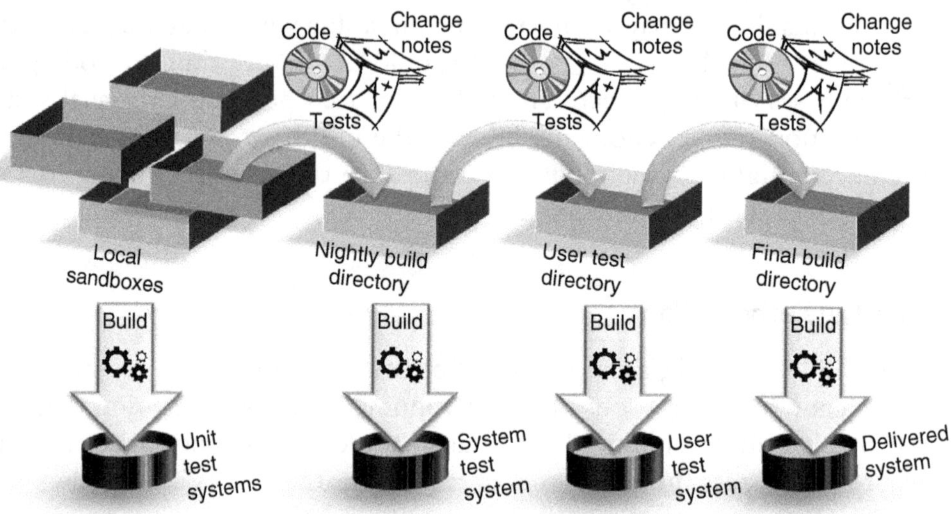

Figure 3. The CBT approach.

remove problems of sending unworkable code to sites and in delays in integration and user testing.

This method allows developers to work in their own sandbox, provides a check that the system can be built and run in an integrated environment, provides a system for user tests before release, and always has a minimally workable system available for distribution.

MOVING THE BALL FORWARD

On every software project it is inevitable that someone makes the statement "Just give me a handful of developers and I could get this done in half the time!" In large development projects everyone seems to spend most of their time in meetings, letting everyone else know what is happening and what the problems are. There often seems to be little effort done "in moving the ball forward" (using sports terminology). In fact, most of the effort seems to be moving the ball backward when user requirements change and new solutions are discovered.

More requirements don't help

There is a method to address these problems. The method can be controversial because it goes against the belief that more requirements are better. Many project managers believe that if only the users could write better requirements then projects would not be late and over budget.

Unfortunately, often users cannot articulate their requirements because these requirements change over time. Requirements change because users have a hard time imagining what they need until they actually see something to evaluation. Users also do not often know what is actually possible and usually only envision products that are faster and cheaper. The methods that address the difficulty in getting true requirements are based on Scrum and the Agile methodology.

- Scrum is an incremental framework for managing projects with rapidly changing or not well understood requirements.
- The Agile software development methodology is a way to provide incremental working elements that end users can evaluate early in the project.

Form a Scrum team

The name Scrum comes from rugby, where teams move the ball forward in small steps as a unit by passing the ball back and forth. In software development it describes a small team working closely with end users to move the project code forward in small steps. Scrum defines a set of practices and predefined roles for running a project through a series of small iterations called sprints.

The methodology is very formal, despite the sporty name. Each sprint is 10–20 working days long. Each sprint has a well-defined goal, and each sprint includes user evaluation at the end of the sprint.

A Scrum team typically has five to nine members. These include a "product owner," who is the user/customer representative and the ScrumMaster, who is responsible for helping the team use the Scrum process and protecting the team from irrelevant outside influences. The team members are identified as either "chickens" or "pigs." The terms come from a joke about a chicken and pig:

> *A pig and a chicken are walking down a road.*
> *The chicken looks at the pig and says,* "Hey, why don't we open a restaurant?"
> *The pig looks back at the chicken and says,* "Good idea, what do you want to call it?"
> *The chicken thinks about it and says,* "Why don't we call it 'Ham and Eggs'?"
> "I don't think so," *says the pig,* "I'd be committed but you'd only be involved."

The product owner, ScrumMaster, and team members are pigs—they are committed to the project and process. They have a personal stake in the success of the project. Managers, other users, and stakeholders are chickens—they are just involved in the process. The chickens want it to succeed, but they don't have a personal stake involved. Both pigs and chickens are required for a Scrum project, but the pigs do the real work.

Becoming Agile

Scrum is often paired with the Agile development methodology. Agile defines a set of software development methodologies based on iterative development, where both requirements and solutions are co-developed through a collaboration process. Agile defines software development aspects that include:

- Continual peer review of developed code. No code is released unless at least two people have read it, the writer and their "buddy."
- Real-time interaction with users to evaluate ideas and implementations. This interaction is not performed using paper-based documents, but is performed using storyboards, mockups, and prototyped code.
- Rapid development cycles that are measured in hours or days.
- Agile teams that work on specific, small, well-defined goals that can be written on a 3 × 5 card. The size of the card limits the scope of the goal.

There are many names for variations of the Agile methodologies, including Extreme Programming, Feature Driven Development, and Dynamic Systems Development. Early adapters of the Agile methods were compared with "cowboy programmers," but the methods are well defined and, when honestly applied, can lead to significant reductions in project time and much higher rates of successful projects.

In February 2001 a group of developers published the *Manifesto for Agile Software Development*. The Agile Alliance, a nonprofit organization, promotes software development according to the manifesto's principles.

Twelve principles underlie the Agile Manifesto (http://www.agilemanifesto.org/principles.html). They are:

- customer satisfaction by rapid delivery of useful software;
- welcome changing requirements, even late in development;
- working software is delivered frequently (weeks rather than months);
- working software is the principal measure of progress;
- sustainable development, able to maintain a constant pace;
- close, daily cooperation between business people and developers;

- face-to-face conversation is the best form of communication (co-location);
- projects are built around motivated individuals, who should be trusted;
- continuous attention to technical excellence and good design;
- simplicity;
- self-organizing teams; and
- regular adaptation to changing circumstances.

If your projects never seem to complete on time and budget, then Scrum and Agile may be your answer. They are well-proven approaches to software projects. Work with your management and executive sponsors so that they understand that moving the ball forward in small steps is better than hoping for a single "miracle" step that will deliver what the customer really wants.

USB THUMB DRIVE DIAGNOSTICS

Next to your compressed air duster and screwdriver in your IT toolkit, you should make room for a USB thumb drive. Most people know these thumb drives, both cheap and plentiful, as simple storage devices, but with some simple work, you can leverage a thumb drive as a powerful diagnostic tool. By loading your drive with the proper software, you can assure that you are ready to root out a variety of computer maladies.

Take note: the advice given here is current at the time of printing; you will want to explore other software options as much of the discussed software must be kept current. Keep an eye out for better options and new software, as well as new problems you might have to address.

What's the problem?

When putting together your diagnostic thumb drive, you should consider the various computer problems you might typically encounter and must be able to solve. The first problem you can address is malicious software. Although good practices and the use of a DMZ should mitigate the risk to many systems, it is not eliminated, especially outside of the DMZ. Computer viruses, spyware, trojans, and rootkits can all infect systems. Therefore, your first goal should be building tools to remedy these problems. There are some software packages that deal with some, but not all, of these attack vectors, so be sure to cover all of your bases when choosing software.

A good basic choice is RootkitRevealer and Ad-Aware (http://www.lavasoft.com). Make sure to keep the programs' definitions up to date on the thumb drive. You will need to do this manually as the programs will not run when the drive isn't plugged in, and you may not be able to update on the computer you are fixing.

For other software ailments, not caused by malicious programs, good diagnostic software can help you track down other problems. The sysinternals package, found at http://technet.microsoft.com/en-us/sysinternals, is a suite of tools that run on Windows systems. There are various diagnostic programs included, such as Process Explorer (a comprehensive task manager) and Process Manager (an event monitor). You can read more about these tools on the website.

When you can't get started

The above tools are useful if you can start up your Windows system. However, if you are unable to start up your system, they will not do you any good. In that case, there's one more solution with which your USB thumb drive can help: You can create a USB boot disk.

Current computer BIOS's allow for booting your computer from an attached USB hard drive. If you are comfortable with Linux, you can easily create a persistent bootable USB drive (you will want a second stick to put this on), which will allow you to boot without using the computer's hard drive or OS.

With a bootable USB drive, you can perform tasks such as memory tests, virus scans using ClamAV (http://www.clamav.net), or retrieve files off of the hard drive. To make a Linux boot drive, you can use Linux Live USB (http://www.linuxliveusb.com), an easy-to-use program that will also download a Linux distribution to load onto your USB drive.

Protect your toolkit

When using the USB drive it is important to take precautions so that your USB drive does not become an attack vector. Except when you are updating your toolkit (such as downloading new anti-virus definitions), keep the read/write switch on the drive as read only. This is especially important if you are using the thumb drive to diagnose a system that you think has been infected. Your do not want to make your IT toolkit part of the problem.

THE FUTURE IS VIRTUAL

Future IT environments will be built on virtualized systems, and this will include the multiple IT systems used in manufacturing. This includes databases, historians, HMIs, schedulers, and even controllers that we use to run our manufacturing operations. Today virtual systems have many forms, from Internet-accessed “cloud”-based systems to company-owned server farms running hypervisor servers.

An old idea made new

Virtual systems are an old idea that has been resurrected with modern technology. During the early days of computing, in the 1960s, there was a concept called time sharing option (TSO) available on IBM’s MVS operating system. It essentially made the single IBM mainframe computer look like multiple IBM mainframe computers each linked to a separate terminal, each running a single application. The modern equivalent takes this to the next level and makes a single server actual like multiple servers at the operating system layer.

There are multiple methods for virtualization, but the one that will be most common in all future server farm environments will be based on the hypervisor concept. A hypervisor is a stripped down operating system that provides a virtual environment for other operating systems. It masks the underlying hardware by providing a complete virtual hardware environment. The virtual environment provides each virtual machine with a number of core processors, memory, disk space, and network access. The hypervisor runs guest operating systems, such as Windows Server 2008, Windows 7, Linux, and even DOS. The most used hypervisor systems are VMWare’s vSphere (a stripped down and customized Linux), Cirtix’s XenServer (also a stripped down and customized Linux), and Microsoft’s Hyper-V (a stripped down and customized MS-Server 2008).

Moore’s law

Moore’s law is the reason that virtualization will become the dominant IT implementation architecture. Moore’s law is an empirical observation made in1965 that the number of components in integrated circuits that can be placed on a chip doubles every two years. Forty-five years later, most experts believe that we still have several more doubling cycles to go before we reach hard physical limits. While the speed of processors has started to level off at about 3 GHz due to limitations of heat dispersion and the speed of light, chip manufacturers have instead added core processors to chips. Four, eight, and sixteen core processors are common, and future systems will have hundreds of core processors.

Memory density has also increased, following Moore’s law, making massive amounts of RAM available for the multicore processors. Disk density and network speeds are also doubling at their own rates, providing terabytes of inexpensive disk storage and Gigabit network access. These advancements mean that a single virtualization server can replace tens of nonvirtualized servers, and this will cause a major change in how we build, buy, use, and manage IT services. With so much computing power available in a single server, the old IT concept of one server per main application does not make sense.

Building a virtual system

One important feature of the hypervisor environment is that it takes only minutes to create a new virtualized server. All of the major hypervisor systems allow you to take a snapshot of an existing system and to use that snapshot to create a new server by copying the snapshot file. The first installation of an operating system takes the normal amount of time, but later installations can be done in minutes. Most virtualized IT departments keep snapshots of multiple operating systems at different patch levels so that they can quickly build any system that an application needs.

The ability to quickly build and deploy new servers means that we can take advantage of the massive computing power available in virtualized systems. We can deploy existing software systems much more rapidly than using physical servers.

The utility model for IT

The IT hardware model is rapidly becoming a utility system model, just like the electrical power model and telephone model. In yesterday's environment it may have taken weeks or months to order a new server. This includes time to obtain approval, check to see if there will be space for the server in the server farm, determine the hardware required by the application, order the hardware, install the operating system, and add it to the corporate network. All this must be completed before the application could even loaded on the server. Compare this business model to ordering a new phone line or an additional electrical outlet. In the utility model the support organization manages the available resources, autonomously adds or deletes hardware to meet needs, and rapidly provides requested services.

The modern IT environment will run in a similar manner. The cost of a virtual server (processer, memory, and disk) in a virtualized environment is only a few percent of the cost of an equivalent physical server. Virtual server costs are so low that they will probably be handled as normal expenses and not as capitalized purchases.

In the future you will decide on the application you need, obtain the required server specifications from the vendor, order a virtual server from the IT department, and install the software in hours or days instead of months. IT departments will manage the virtual servers, expand the server farm to meet expected needs, and decommission virtual servers (free their resources) when the applications are no longer needed.

WHY TO VIRTUALIZE

Virtual systems are rapidly invading existing server environments. Massive racks of servers in air conditioned and environmentally protected computer rooms are being replaced by smaller racks of servers with much smaller power and cooling requirements, yet with more performance, memory, and disk space.

Even manufacturing IT systems, which are often housed in small server rooms closer to the shop floor, are being replaced with remote virtual servers, often on the same site but at a central location. These changes will significantly simplify many of the problems seen today in manufacturing systems, such as upgrade methods, backup methods, and standby methods.

The economics of virtualization

While the economics of virtualization will drive IT departments to virtual systems, the advantages to manufacturing mean that manufacturing IT departments should actively encourage and support virtualization.

Economics are driving companies to move to virtualization. A single redundant host server with 32 GB or 64 GB of memory, a few Gigabit network connections, and supported by a multiple-terabyte RAID Storage Area Network (SAN) disk server can cost 70–80% less than an equivalent-capability small server farm. Load monitoring of manufacturing IT systems show that a single host server can handle anywhere from 4 to 12 virtual machines (VMs) running typical manufacturing IT applications.

Measured savings on a typical system, as presented at the December 2010 ISA/ISPE conference in Research Triangle Park, NC, documented capital equipment savings of 72% ($50,400 reduced to $13,000), yearly direct energy savings of 42% (4,788 W to 700 W), and BTU cooling reductions of 85% (16,000 BTU to 2334 BTU). In one implementation, a single server with 35 GB memory and 2 TB of disk replaced four historian servers, one I/O server, six dedicated workstations, one backup server, one print server, one domain controller, one SharePoint server, and one SMTP server, and still provided for a 50% growth capability. Because of these significant savings, all servers in most facilities will quickly move to a virtual environment.

Virtualization of updates

The move to virtualization also brings other benefits, especially in the area of updates. One common problem in industrial systems is managing all of the updates from vendors. A typical site may have 20–80 different applications, usually from multiple vendors, and many at different operating system versions. Managing changes to the applications and the operating systems is a full-time job, and it is especially difficult because changes to production systems must be planned much in advance so that the upgrade does not stop production. Virtualization provides a good solution to this problem.

Most virtualization systems provide the capability to make a "snapshot" of a system. This is a copy of the current system that can be used to rollback if changes made to the system must be undone or to make copies of the current VM and clone it onto another VM. The ability to clone a system can be used in your upgrades process. It gives you a method to make changes to a copy of the current environment, run tests to make sure the changes work and don't "break" anything, and then copy the changed version back to the running version. This allows you to make the changes off-line, without affecting production, but ensuring that you are working on the exact current configuration. To make this method work you will need to set up a complete separate environment for the cloned copies.

This environment must include all of the peripheral services; I/O services (such as OPC servers), print services, domain services, security services, license services, and file share services. While you may be upgrading only a single VM, you will probably need copies of all the VMs within your system to re-create the production environment. The cloned copy must be installed and tested on a completely separate network, because you do not want to have to change the name of the servers for your tests, and this requires a separate domain controller. In some environments you may be able to run the cloned copies on the same VM host and configure the clone VMs to run within their own internal self-contained network.

The cloned VMs will have no access outside the self-contained network, and no one can see into that network. Once you have performed the update and verified the update with your local tests, you can take a snapshot of the updated system and copy that to the production environment. You will still have to halt production to stop the production servers, install the copies, and restart the servers, but this will be significantly faster than installing the updates on the production system and running your tests.

Hardware and energy economics will drive IT departments to virtualization, but the additional functionality provided by snapshots and rapid cloning of systems are valuable features for manufacturing. This additional functionality will reduce production down time and reduce the risk from upgrades that break the current system. Manufacturing IT departments should learn the VM technologies and capabilities and actively encourage and support your company's move to virtualization.

VIRTUALIZE NOW

The ability for virtualization to simplify system upgrades is a major reason to move to virtualization. There are at least two additional reasons to virtualize, but three major issues to address.

Patch Tuesday

The first Tuesday of every month is a significant date for most IT departments. It is the day that Microsoft releases patches. On the first Wednesday of each month, an army of IT workers at vendor companies start testing the patches on their installed systems and applications. With luck and no compatibility problems, the vendors release advice to their customers about the patches. Meanwhile, end-user companies start performing tests on their systems and applications before they rollout the patches to their production systems. A major problem with this model is the requirement to keep exact copies of production systems available for the patch testing. When changes or updates are made to the production systems, the same changes must be identically applied to the patch test systems. This doubles the hardware cost of a system and doubles the maintenance costs.

Virtualization simplifies this through the use of snapshots. Snapshots are an exact copy of a system's memory, databases, and installed applications. A snapshot can be loaded on a Virtual Machine to give you an exact copy to use in patch testing. There are two common methods to use: snapshot as rollback/backup and snapshot as test copy. The rollback/backup model is used when short-term unavailability of a system is acceptable. A snapshot is taken of the production systems before the patch is installed, then the patch is installed, the system is restarted (if necessary), and is quickly tested. If no problems occur, then production continues with the patched system. If problems occur, then the system is rolled back to the snapshot.

The "snapshot as test copy" method is used when production system downtime is not acceptable and a patch reboot is not needed. A snapshot is taken of the production system and then installed in a test VM environment. The patch is applied to the test environment and tested. If the patched system passes the acceptance tests, then the patch is applied to the production system.

Redeploy on new hardware

Another advantage of visualization is the ability to move a VM from one host server to a new host server without shutting it down. This functionality provides a method to upgrade hardware without shutting down applications and impacting production. This also provides a method to increase the CPU power available to an application on an on-demand basis. Not all hypervisor VM systems provide this functionality, so it is important to see what your virtualization system provides. It is also important to test this feature prior to implementing it in production because not all hypervisors handle the switchover in the same manner.

Virtualization issues

There are three issues in using VMs in manufacturing IT systems, unique peripherals, industrial networks, and vendor support. VM systems have two unique characteristics: all communication must be through an Ethernet network connection, and no special devices can be used. This means that most USB devices, printers, and license dongles must be connected through Ethernet. Also industrial networks must be connected through Ethernet, special video cards cannot be used, and special keyboards cannot be used. Some Virtualization servers will allow you to dedicate a USB port or communication port to a VM, but then the ability to move the VM to different hosts on an on-demand basis is lost.

There are off-the-shelf solutions for simulating USB devices, printers, and license dongles across Ethernet, but you will have to test the required drivers and network-connected devices in your environment. Industrial networks are a bigger problem. Any Ethernet TCP/IP-based protocol will probably work in the VM environment, but industrial networks that require proprietary network cards and non-10/100/1000 MB Ethernet cabling will probably not work in a VM. Fortunately, there are many Ethernet-to-industrial network convertors available. For example, there are Ethernet-to-Modus devices that allow 30-year-old Modbus connectivity to VMs. The newer OPC protocols (OPC-UA and OPC.NET) are Ethernet compatible and usable in the VM environment. OPC-UA is suitable for both tag-based data and complex data, and OPC.NET is suitable for tag-based data. While there may be some concern about the use of Ethernet because of the nondeterministic nature of the protocol, the availability of 1000 MB networks have essentially removed this as a valid concern.

Vendor support

The final major issue is vendor support. Check with your vendors to understand their support policy for virtualization. Many have some support, but may require that you run unvirtualized if a problem develops that they need to duplicate. However, once you move to virtualization, backing out for troubleshooting may not be possible. The little secret that many vendors have is that most are using virtualization in their own development and test environment. Virtualization is a technology and solution that is good for manufacturing and helps us apply IT solutions to more industrial problems.

Cyber Security

Entire volumes have been written about computer security but few directly address the specific requirements of manufacturing systems. This chapter describes the unique characteristics of manufacturing IT systems and how security can be applied to these systems. You are introduced to the terminology and concepts that are critical to keeping manufacturing systems safe from intentional and unintentional attacks.

The first section defines some of the unique security concerns for manufacturing systems and some simple initial steps you can take to separate mission-critical production systems from other corporate systems. One step is the construction of a demilitarized zone (DMZ) that lies between mission-critical production systems and less critical systems. This concept has been effectively applied to reduce the risk of viruses and worms on manufacturing systems impacting business systems, and virus and worm attacks on business systems impacting manufacturing systems.

Determining how much effort and money to spend on security is often a hard problem, because it takes money away from projects that increase efficiency or productivity. However, money spent on security is a good investment if the proper amount is allocated. This chapter defines some rules of thumb you can use for estimating the cost of security.

Installing antivirus systems on production systems and implementing password rules always raise contentious issues about the balance of security, performance, and usability. The last two sections in this chapter list some of the best practices for anti-virus systems and password rules for manufacturing systems. This information helps you develop good security policies for your installations.

VIRUS AND HACKERS AND WORMS—OH MY!

Computer viruses and worms are always big topics in the IT world. Worms and viruses have infected company LANs and even shut down businesses. While these concerns were already important in the IT environment, they had not been as important in the control system environment. With the increasing use of standard Ethernet, common networks, and Microsoft operating systems in control systems, infection concerns now have to be considered in control system design and support. As proof of this, many companies have had to stop production because of the recent attacks or because of the actions taken in response to the attacks.

ISA 99

Part of the modern control system engineer's skill set must include knowledge of how to protect networked control systems. The ISA TR99.01 Technical Report on Security Technologies for Manufacturing and Control systems is a good place to read about the technologies that you would need to apply.

IT systems generally follow three rules for protection: defend at the edges, detect in the interior, and protect at each system. Defending at the edges means stopping viruses and worms from entering the local network. This includes firewalls, e-mail scanners, closing unused ports, and requiring security access control on any communication through the firewall.

Detecting in the interior is scanning of network traffic for suspect and abnormal activity. Detection may also involve scanning server systems to make sure that the approved applications, and only the approved applications, are running. Protecting each system uses virus protection software and personnel firewalls or each system. These same rules can be applied to networked control systems with one important exception. The exception is "protecting at each system." Virus protection software requires continual updates of virus and worm electronic signatures. This usually involves downloading identification files and often requires installing software updates.

Unfortunately, it is unacceptable to make these changes without extensive testing and revalidation on validated or critical control systems. Updates can occur several times per week, but testing and validation can take weeks so it is nearly impossible to have current up-to date virus protection software on validated or critical control systems.

Defend better at the edge and in the interior

Since we cannot follow the third rule on many networked control systems, the first two rules should be strengthened to take up the load. We can strengthen the first rule by adding firewalls between the control system networks and the rest of the corporate networks. Unprotected control systems are prime targets for infection, and they need multiple layers of protection. Control system networks that connect directly to other business system networks are at risk from viruses and worms and put other corporate systems at risk.

Firewalls with limited ports provide one level of protection. The firewalls should be two way, in addition to protecting control systems from infection by corporate systems, it must protect corporate systems from the control systems. Access control routers can also be added to augment the firewall protection. Access control routers allow only specified systems on one side to access systems on the other side. The control system network can also be designed as a Virtual Local Area Network (VLAN) using intelligent switches. The VLAN isolates traffic on the VLAN from other LANs, providing an additional measure of protection against broadcast storms and other denial of service (DOS) attacks.

Detection within the control system network should also be applied. This includes using Intrusion Detection Systems on the local VLAN. One easy method is to set up an infection target system. This is a system running internal monitoring software to detect infection with all incoming ports open, and all output ports blocked. This system acts like the "canary in the mineshaft"; it will be the first system to detect a problem.

Use commercial off-the-shelf solutions

The cost of adding infection protection to control systems is small and is available with off-the-shelf software. Yet, the benefit of infection protection is immense, because infected systems can shut down production, damage products, and place personnel at risk. Control system professionals need to understand the technologies of infection protection and must work with IT departments to implement secure interfaces between the control networks and the corporate networks.

BAR THE WINDOWS AND TIE DOWN YOUR HAT

Every month we are reminded by the arrival of the latest Microsoft patches that it is time to lock the doors and bar the windows in our factories. Recent electronic attacks have shown that Internet-based attacks have reached a new level of sophistication. Experts, if we have to use the term, are involved in cybercrime using root kits, DNS spoofing, DNS redirecting, and zombie computers. The patches are not just related to Microsoft's products, there are security holes in almost all software, including Linux, the Firefox browser, Oracle databases, Cisco servers, and Sun's Java Web Proxy Server. The USA Department of Energy Computer Incident Response Capability at www.doecirc.energy.gov has lists of current vulnerabilities.

Up to now most attacks have been in the financial industry for economic gain, but cybercrime in manufacturing and cyber terrorism will not be far behind. Intellectual property theft in manufacturing is on the raise, especially from countries that do not have the strong IP protection laws of the USA and Europe. Cyber-attacks are becoming easier to launch and harder to detect. "Root-kit" hacks for taking over computers are readily available on the web and are being used to launch internal attacks on corporate information.

Robust protection required

In the face of the growing threats it is important to put robust protection for manufacturing assets in place. A February 2005 report by the British organizations NISCC, the National Infrastructure Security Co-ordination Centre (now the Center for the Protection of National Infrastructure), titled "Firewall Deployment for SCADA and Process Control Networks" provides invaluable help for system protection. The report is available at http://www.cpni.gov.uk/documents/publications/2005/2005022-gpg_scada_firewall.pdf. The report discusses several different network configurations using routers and firewalls, illustrates the strengths and weaknesses of each approach, and rules for firewall configuration. While every process control system may not need the highest level of protection defined, every system should have a minimum level of protection. The report authors consider that non-firewall-based systems will generally not provide suitable isolation between the process control networks and enterprise networks. These include systems using Virtual LANs (VLANs) and routers to separate the process control and enterprise networks.

The DMZ approach

The best practice recommendation in the report is a three-zone approach, depicted in Figure 4, using dual firewalls with a demilitarized zone (DMZ) between the enterprise network and the process control network. One firewall protects the HMI, PLC, and DCS systems from undesired outside influence. Above the process control firewall is a DMZ that contains data historians and other databases for shared information. A second firewall separates the DMZ from the enterprise network.

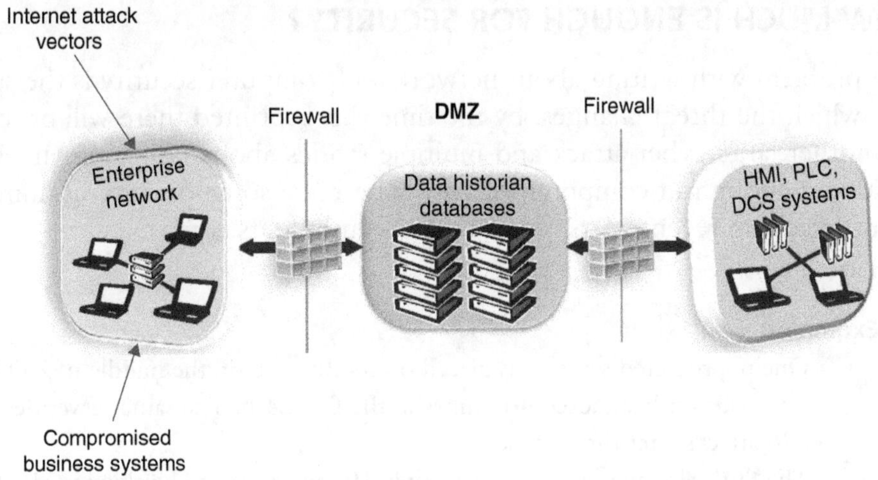

Figure 4. The DMZ approach.

There is no direct access between the two networks because all communication is through applications running in the DMZ servers. The DMZ applications are typically SQL databases, plant data historians, and scheduling and reporting applications. This provides the most secure environment by limiting the possibility of attacks from the enterprise compromising the reliability and timing of the process control network, and protects the enterprise network from infection by compromised automation systems. The enterprise firewall blocks arbitrary packets from getting into the DMZ or process control network. The process control firewall prevents communication from a compromised server from entering the process control network, and prevents process control network traffic from impacting the DMZ server network. Although not discussed in the NISCC report, MES and batch execution applications should probably sit on the process control network, with the MES databases and batch logging databases in the DMZ.

The three-zone approach has the added advantage of having limiting the number of applications in the DMZ that need to be kept up to date with patches and virus protection. Since these applications will often be commercial products that are not specific to automation, they are easier to test and validate for patches and virus updates.

The NISCC report provides a good definition of industry best practice for IT security in manufacturing systems and should be required reading for all control engineers.

HOW MUCH IS ENOUGH FOR SECURITY?

One problem with writing about network and computer security is the speed with which the threat changes; by the time this is printed there will probably be another large cyber-attack and multiple stories about how companies are not doing enough in computer security. There are some disturbing numbers about attacks in recent years reported on SecurityStats.com.

Examples:

- One unprotected server was placed on the Internet in the middle of 2003 and was maliciously attacked 467 times in the first 24 h. The same server detected 626 attacks after three weeks.
- The SQL Slammer work required only 10 min to spread worldwide, doubling in size every 8.5 s, much faster than any possible manual intervention.
- Remediation costs of the MS Blaster worm were almost half a million dollars per company, with large companies reporting losses in the millions.
- At its peak, one in twelve e-mail messages on the Internet were sent by the MyDoom virus.
- Stuxnet, a Windows trojan, was detected in June 2010. It is the first worm to specifically attack control systems. It raised the risk of infection because of the multiple attack methods and ability to damage specifically targeted production equipment.

PC viruses cost businesses billions of dollars each year in lost time and recovery costs. Manufacturing operation systems are not immune from these problems, privately multiple companies have reported losses of millions in production due to virus and worm attacks in 2003 and 2004. The important message is that virus and worm attacks have shut down production in inadequately protected operational systems. This begs the question, what is the right amount to spend on security and the related network infrastructure?

How big is the problem?

Usually a company's firewalls and security devices protect the corporate intranet and the operations and automation networks. However, it is still advisable to separate operations and automation networks from corporate intranet using firewalls, VLANs, or physical separation. Automation and operation systems are often mission-critical systems. This means they must remain operational for production to continue. Unfortunately, these systems often are not running current virus protection and current patches. This is not for lack of trying, but in 2010 Microsoft released 106 patches across all products, or about two patches per week. Virus protection updates come almost daily because new viruses and worms are released daily by cyber-vandals. In the face of this onslaught of attacks and protection measures, it is important to know how much should you be spending on operations and automation security.

What's the cost?

In general, on the basis of several public surveys, security hardware, security software, and security personnel seem to be about 4% of IT budgets. Some industries, which have mission-critical IT infrastructures, such as financial organizations and universities, spend more, averaging about 7% of their IT budget and in a couple of cases up to 20% of their IT budget. An additional 7% is being spent on network infrastructure with some of that money going to security issues. The META Group estimates the average security investment will peak at 8–12% of IT budgets in the United States by 2006. The security portion of IT budgets is split about one-third on security hardware (firewalls, intrusion detection systems, e-mail scanners, etc.), one-third on security software, and one-third on security personnel.

On the basis of industry standards for mission-critical applications, the average manufacturing IT organization should be spending about 5–10% of their manufacturing IT budget on security. This is a small number and easy to forget or ignore in capital projects and yearly budgets, but they are important. These items are similar to insurance, they represent a cost until they are needed, then they are definitely worth the money spent. This insurance is vital because the latest generation of worms and viruses use picture files (JPG), sound files, and other previously safe data files for infection. Unfortunately, in today's cyber-terrorist, cyber-criminal, and cyber-vandal world, we have to pay a price for protection, and it is important to know how much is enough for security.

ANTI-VIRUS PROTECTION, IS IT FOR YOU?

An important question always comes up when installing control system applications: "Should it run anti-virus software?" The answer is usually an easy "yes" when the applications are commercial databases, historian systems, or other non-real-time control applications. However, if the applications involve real-time control or guaranteed response times, such as HMIs, DCS systems, or PC-based control systems, then the answer is not so easy. There is a valid concern in control applications that antivirus software may seriously impact performance, disrupt production, and void control system vendor support contracts. NIST (The National Institute of Standards and Technology), Sandia National Laboratories, manufacturing companies, and control system vendors recently conducted a study on the impact of antivirus software on industrial control systems. The study not only confirmed some fears but also point to a path forward for the additional protection offered by antivirus software.

Bad news and good news

First the bad news: manual scans and scheduled full system scans will take up all of the available CPU cycles, pushing utilization to 100% for extended periods of time. This can seriously impact application performance. Reducing the priority of the scan reduces, but does not eliminate the performance impact and also increases the period of instability. Signature updates, where new virus signatures and sometimes new scan engines are downloaded, can also take up all of the available CPU cycles, but usually for less time than a full system scan. Even worse, signature updates may also require reboots.

The good news is that active scanning, which is the scanning of executables and libraries prior to execution, has minimal impact on industrial control systems.

The last line of defense

Anti-virus software should be only one layer in a multilayered defense. It is usually the last line of defense before your system is compromised. Anti-virus software supplements other layers, such as firewalls between business and control networks, separate user authentication controls for control networks, separate network access controls for control networks, network-based intrusion detection systems, and strict control of installed applications.

Antivirus software can be applied to industrial control applications but several rules should be followed.

First, validate that the anti-virus software works with the system applications by testing in a separate system, and validate that the software will not void your vendor support agreement. System owners may have to test several antivirus solutions and may need separate subscriptions for control system applications.

Second, disable scheduled full system scans. This will prevent an inadvertent scan at an inconvenient time, such as in the middle of the night during a rush order job. Initiate all full system scans manually on a regular basis at know down times or when the impact of the scan will not affect safety or quality.

Third, use a local virus definition server and do not directly allow virus signature updates from the antivirus vendor's systems. This serves three purposes: it allows the system administrators to plan and schedule the updates at safe times, it removes a direct link from the control system to the Internet, and it allows the system owner to test the impact of the new signature or scan engine prior to distribution.

Fourth, keep the application servers clean. This means uninstalling unnecessary applications, especially those that may come preinstalled on commercial servers. This also serves several purposes. It reduces the number of files that must be scanned, removes hidden direct links between the application server and the Internet, and reduces the applications that can be infected.

Antivirus software is often the last line of defense in secure systems and applying these rules allows antivirus software to be run on control systems with minimal performance impact. However, support and diligence are needed to perform full system scans and signature updates under manual control.

PASSWORD PROTECTED?

Computer passwords can serve as a useful layer in your overall security. As ubiquitous as they are, however, passwords are often weakened by everyone's desire to have an easy-to-remember password. The desire for simplicity and easy-to-remember passwords is in direct conflict with effective security. Many companies have standard policies for constructing passwords, but not all manufacturing systems have methods in place to enforce the policies. In these cases users often create passwords that are vulnerable to cracking and they greatly reduce the password's security effectiveness. To keep this from happening, try to keep the following password suggestions in mind:

1. Don't use dictionary words. The primary method of "cracking" or defeating a password is to use a dictionary file—a file containing words from the dictionary, as well as words followed by simple suffixes (e.g., freedom17). Names are also on the list, so avoid names of pets, the names of loved ones, or sworn mortal enemies.
2. Make sure your password is "long enough" with a good set of characters. If a dictionary attack fails to crack a password, there are other options available. One sure fire method is a brute force attack, wherein every possible combination of characters is tested in sequence. In 2010, the Georgia Tech Research Institute developed a method to use graphic processors (GPUs) to help crack passwords. GPUs are the computing element in most graphic boards. The parallel processing capability that GPUs have means that trillions of passwords

can be checked per second in a high-performance graphics cards. By looking at the math behind all possible passwords, you can see how to delay the success of such an attack. The possible number of combinations of *N* different characters of a password of length *L* is *N^L*: increasing either the number characters you can use or the password length can increase the difficulty of cracking the password. For example, if a hacker can enter 1 trillion passwords a second, it would take just 5 min to brute force crack a password consisting of lower case letters that is six characters long. However, if a password uses all available symbols (~95 on a keyboard) and is 14 digits long, it would take over 100 million years to crack. Passwords should be at least 10 characters long, use upper and lower case letters, include numbers, and at least one special character, such as $, %, &. *, (,) , or !. Some systems will limit the number of special characters you can use, in this case use passwords at least 12 characters long.

3. If possible, don't reuse passwords. If you use the same password, or even the same username and password combinations for multiple systems, a compromise to one system will compromise your account on all the systems. The password you use for a random website log in could become part of someone's password dictionary file if the site is compromised. This not an idle speculation, it has happened many times: in 2010, Gawker Media had a security breech in which the passwords used by commenters on several websites were exposed, as well as their e-mail addresses. You should use a different password for every different log in that you have. This is a difficult rule to follow; in fact, it is impossible for most people to follow. Most computer users will work on multiple systems and keeping track of dozens of passwords will tax anyone's memory. Writing the passwords down in a "little black book" is also not secure, but if you do keep the book locked up when not in use. Another option is to use a password manager. This is a program that will keep an encrypted database of your user accounts and passwords and can even generate truly random usernames and passwords for each different system you use.
4. If you must reuse passwords because of your system's architecture, then use internal passwords that are different from external passwords. If your systems are to only be used within a secure environment, don't compromise it by using a password that you use to log on eBay or Amazon. Make sure that the passwords you use for work are different from those you use for your own personal use. That is if your company is compromised, then your personal accounts are not compromised. Likewise, if your personal username and passwords are compromised, this will not compromise your company's systems.

If you keep these guidelines in mind when creating passwords, you can make sure that they remain as functional a security measure as possible. Although they do not make it impossible for someone to get into the system, they greatly increase the chance of the systems withstanding someone's attempt at access.

Certain methods can circumvent even strong passwords, such as leaving the password on a post-it note on the monitor or booting the system from a Live CD, so remember that physical security is also an important concern. If you wish to read more about password security, Microsoft and Google offer more advice on their websites:

http://www.microsoft.com/security/online-privacy/passwords-create.aspx

and

https://www.google.com/accounts/PasswordHelp

Databases

Databases are everywhere. The collected information in databases has been calculated to exceed 1 zettabyte (1,000,000,000,000,000,000,000 bytes) by 2015. Databases are a key technology for IT because they provide the way to store and rapidly access incredible amounts of information. It is important for any manufacturing professional to understand something about database technologies, and this chapter presents some of the basic concepts of databases. The first section in this chapter discusses putting the "Information" back into IT.

Sometimes all you need is a database for a project, and the next section in this chapter discusses when a database may provide all of the functionality you need for small manufacturing IT projects.

Databases are based on mathematics, and the third section in this chapter discusses the mathematics of information in relational databases. You should understand the solid basis for database design and use so you can design effective manufacturing databases.

The last section in this chapter defines some practical but simple rules for designing manufacturing databases.

LET'S PUT INFORMATION BACK INTO IT

IT means Information Technology, but often IT departments focus on the technology side and do not give the information side equal attention. The information side of IT deals with the form and format of data and with the data's "metadata." Metadata is data that is used to explain and describe other data.

Example:

- One piece of data may be an employee ID called "EmployeeID" that can have a specific value, such as 1999000023. The metadata for EmployeeID defines the type of the data (a 10-character numeric string) and any rules the data must follow (the EmployeeID must start with the 4-digit year of first employment, and no two EmployeeIDs may be the same).

IT groups manage information by coordinating and controlling values of the data and ensuring that all uses of the data conform to the metadata rules. Coordinating and controlling this information across several different applications can be difficult and is often the hardest part of application integration. In the manufacturing IT domain, one common difficulty is managing equipment tag names because every application seems to have its own rules and restrictions, and these can often be in direct conflict.

Master data and transient data

Shared information is usually separated into two categories: master data and transient data. Master data has several characteristics; it describes business objects that are important to multiple business processes, key parts of the business object must be shared across multiple applications, the complete business object is not completely described by a single application (different applications maintain different elements), the business object can be uniquely identified, and the data are relatively stable, often with a lifetime measured in years.

In manufacturing systems, master data often include material master definitions, physical asset identification, equipment classes, people, and personnel roles. Other shared data are called transient or transactional data and it changes relatively often and can have a limited lifetime. The line between master data and transient data is fuzzy sometimes leading to confusion in deciding what data are master data.

Example:

- A physical asset definition could be master data, but the maintenance status of the equipment could be transient data.

Master Data Management is a process

IT departments have developed a method for coordinating and controlling master data called Master Data Management (MDM). Unfortunately, the acronym MDM has become an often misused and misunderstood marketing term.

MDM is not a technology. It is a collection of management processes that can be supported by data management tools. Don't be swayed by MDM vendors into believing that a single product will solve your data management problems. Without well-defined management processes, all an MDM tool will do is cause confusion and conflict.

Management processes must be defined to handle adding new master data elements, deciding when to remove master data, enforcing the rules for the use of master data, ensuring that no alternate uses and definitions creep into applications, checking all shared data to see what should become master data, and checking all shared data to see if it is already defined as master data (usually with a different name).

Companies that have successfully implemented MDM have set up dedicated MDM process teams to handle master data management processes. The teams are given sufficient authority to require all projects to correctly use master data. The MDM process teams do not look at any specific technology because other groups are responsible for specific MDM tools.

Master Data Management for information management

MDM is a critical element for focusing on information and for building effective application integration, especially for manufacturing to business system integration.

If your company has an MDM team, then make sure that the MDM team understands manufacturing's master data requirements and that manufacturing master data processes are included as part of overall MDM processes.

SOMETIMES A DATABASE IS ENOUGH

With all of the new information technologies available, it is important to remember that often the simplest solution to a problem is the best solution. Sometimes projects start with a design based on complex web-based solutions and dedicated software applications, but can actually be implemented using simple databases. Databases often provide workable and robust solutions, and knowledge of database technology is a vital tool for the modern control professional.

There are several types of databases: relational, network, hierarchical, and object based. However, today the most prevalent are relational databases using Structured Query Language (SQL). In addition to commercially available SQL databases from Microsoft, IBM, Oracle, and others, there are several open-source SQL databases, such as PostgreSQL (www.postgresql.org), Firebird (www.firebirdsql.org), and MySQL (www.mysql.com).

Relational databases

Relational databases are an application of mathematical set theory and in 1970 were described in the paper "A Relational Model of Data for Large Shared Data Banks," by Dr. E. F. Codd, in Communications of the ACM (Association of

Computer Machinery). They are formally defined in ISO/IEC 9075:1992, Database Language SQL, and ANSI X3.135-1992, but usually referred to as SQL-92.

There have been extensions to SQL defined in 1999 and 2003, but the basic relational model has remained the same. In a relational database, data are collected into tables and a table is called a relation. Each table contains columns, called attributes, and rows, called tuples. Each row represents an object, and each column contains an attribute value for the object. Tables are related by values of attributes (hence the word relational).

Example:

- Two tables may have an equipment serial number attribute, and when two serial numbers are the same in the two tables, then the rows in the two tables have a relationship.

This example results in duplication of information, but a process called "normalization" is used during database design to ensure that the tables are effectively defined for minimal data duplication.

Database administrators

When you need to build a database for a manufacturing problem, you will have to get involved with yet another IT organization, the corporate database administrators. Most companies have an IT group that is responsible for managing the structure of corporate databases. They have the responsibility for making sure that data are consistently defined, that appropriate security is added to data creation and access, that the corporate rules for management of data are followed, and that data duplication is minimized. If your database will be part of the corporate system, you will have to justify its existence to the database group satisfaction.

Sometimes much of the data needed for a manufacturing IT project are already available in corporate databases.

Examples:

- There can be an asset management database with equipment definitions that can used to identify production equipment.
- There may be inventory databases with material lot IDs and locations.

Find the database guru

When corporate data already exist, it should be used as part of your project. Usually there is a database "guru" in the IT department who knows about all

Table 3. Database Example

Sales Database					
ProductKey	ShipperKey	ItemQuantity	ItemCode	Discount	CustomerKey
1101	0092	300	200A	20%	99876
1102	0092	250	U665	10%	99876
1101	0085	2000	200B	5%	87762
1902	0085	1	200AX	0%	7566

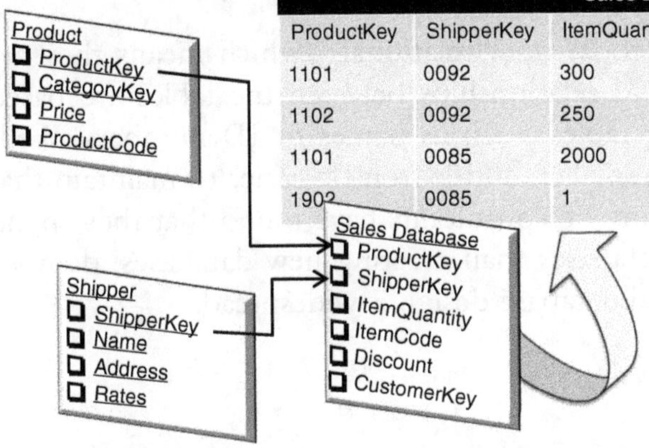

of the corporate databases, and is an invaluable resource for finding data that already exist.

It is also a good idea to work with the corporate database administrators for local projects. They have a lot of practical experience that can be applied to Manufacturing IT databases. They understand databases, but you understand the application. You will have to describe the data and explain its meaning. It is vital to be clear about the name and meaning of tables and attributes, and to document the names and meaning in the appropriate corporate documents. This prevents others from misusing your data and you from misusing others' data. The database administrators understand normalization and will help you design efficient database structures.

Learn SQL

However, talking with them will require an understanding of SQL. A modern control professional should have at least a basic understanding of SQL and relational databases so they can apply simple solutions to simple problems. Table 3 illustrates a common database structure.

LEARNING THE MATHEMATICS OF INFORMATION

Information is the important element of manufacturing IT, and one skill that every well-rounded control engineer should have is knowledge of the mathematics of information, specifically the mathematics of relational databases. Most control engineers will eventually have to build or specify a database to hold instrument data, analysis data, or production reporting data. This is the point when understanding the underlying structure and mathematics of databases becomes important.

The majority of databases today are "relational," which means they are made up of multiple tables, and relationships between the tables are based on common values within the tables, such as names or ID numbers. Well-structured databases are faster, smaller, and extremely easier to maintain that badly structured databases. Many IT consultants have found that they spend much more time fixing bad databases than creating new databases, demonstrating that the problem of bad database design is widespread.

Normalize that data

Well-structured databases are "normalized," but the rules of normalization are often not known or consistently applied. The purpose of normalization is to eliminate redundant information, making it easier to create, read, update, and delete database information. The better normalized is a database, the less special application code or special use rules are required.

Un-normalized databases will often have special use rules, such as "make sure you delete all of the rows in table 'A' that have the specified key value." Normalizing a database is also a key process in ensuring that the database accurately meets its requirements. Normalization cannot be performed when the requirements are fuzzy or uncertain.

It's just rows and columns

Relational database tables are made up of rows and columns. The column's names are called attributes. The individual rows in a database are called tuples, a mathematical term for an ordered list of objects. One or more attributes in a row make up the "key" to the tuple. The primary rule for relational databases is that no two rows can have the same key attribute values. This sometimes requires the creation of an additional attribute to hold a unique ID.

First Normal Form

The first rule of normalization is the removal of repeating attribute groups and creation of a new table to hold the repeating data. Repeating groups are elements that can have multiple values, such as a book that can have multiple authors. In this case a new table would be created for the repeating group, which has a single row for each book/author pair. This is called "First Normal Form."

Second Normal Form

Second Normal Form is obtained by finding attributes that depend on only part of a multifield key and moving them to a separate table.

Example:

- In a book/author table there could be an attribute for the book's abstract, these would be the same for all rows for the same book and should be moved to a separate table.

Third Normal Form

Third Normal Form is obtained by eliminating columns that are not dependent on the key attributes.

Example:

- In a book/author table that could be an attribute that contains the author's company name. In Third Normal Form an author/company table would be created.

Most well-designed databases are at least Third Normal Form. There are additional rules and normalized forms, but these are often not needed for small databases. When you help either design or review manufacturing databases, check that the database is at least Third Normal Form.

Database structures and the rules for normalization are usually taught only in college-level Software Engineering or Computer Science and most engineers never receive formal training. If you want to expand your skill set in the mathematics of information, take a class at your local college or community college in database theory and structure. It will provide you with valuable IT skills.

PRACTICAL MANUFACTURING DATABASES

It seems that the mantra of web services is drowning out simple database applications. Database applications are an inexpensive solution when you do not have an MES system or cannot justify an MES system. Typical database applications involve recording an event, a person, a place, a material, and a time.

Example:

- Take, for example, an operator recording pallets loaded into a loading station. A database application with bar-code input would allow the operator to scan his badge, scan the pallet, scan the loading station, and record the information and time into a database. The database could then be used to support tracking material flows through the plant.

Labels and samples

Another problem is keeping track of production artifacts, such as test labels and samples, as they move across desks for review and sign-off. The application would allow an engineer or operator to scan her badge, the location, and the folder containing the artifact. This allows tracking of the artifacts as they move through the sign-off process.

Simple guidelines

Here are some simple guidelines for database applications:

1. Keep the functionality small and stick to problems that are transactional in nature and not real time or procedural. Transactional problems rarely need more than create, report, update, and delete functionality.
2. Use an SQL-based database. This provides scalability and long-term maintainability of the application. Many manufacturing applications will have a long lifetime and only a standards-based database will provide long-term supportability. SQL databases also have standard reporting and display tools, reducing the application code.
3. Use a database server. While small applications can often be run on a single computer, locating the database on a database server provides a supportable solution. The application's database may be just one of many databases on the server, but the server will have backup support, RAID (redundant) disks, and redundant power supplies.
4. Use integrity constraints in the database. Most SQL databases support some form of FOREIGN KEY, NOT NULL, UNIQUE, and range-checking functions. These functions ensure that only valid data are added to the database and provide checks that do not have to be included in code.
5. Create tables without integrity constraints to hold "abnormal" situations. If an error occurs in normal operation, the operator will often not have enough information to solve the problem. Create "abnormal" tables that have same structure as the normal tables but without the integrity constraints, and store the abnormal cases in these tables. Use this information for later analysis and correction of problems.

6. Use transactional control. Most SQL databases provide some form of transaction control. This allows multiple changes to the database that either all succeed, or, if any one fails, then all rollback. This further reduces the application's error handling code.
7. Use stored procedures. Stored procedures are SQL statements that are executed on database events, such as a change to a data field or deletion of a record. Stored procedures can significantly reduce the application code.
8. Provide an option so that the application creates the database. Testing and rollout are much easier if the application can re-create its database, including the stored procedures. It is easier to keep track of a single application source file than multiple files and versions.
9. Don't hard code database paths, instead use the OS registry database to find environment information. Hard codes paths, such as "C:\My Database," provide an execution constraint that is hard to maintain. Relocating the database without recompiling the application is more supportable.
10. Use Visual Basic or an equivalent to reduce the code you must write and to provide a good user interface. There are many free or low-cost, high-level development environments that can reduce the coding effort and make a database application a cost-effective solution.

Simple guidelines for lasting applications

These simple guidelines have helped me develop long-lasting database applications when an MES system is not available or cost justified.

Regulations

Manufacturing is a highly regulated industry, overseen by multiple regulatory authorities and an increasing number of regulations. Each industry is often covered by unique regulations, but there are some regulations that are common to all manufacturing types. This chapter discusses some of these most common manufacturing regulations that you may encounter and some IT systems that can help reduce the time and effort required to comply with the regulations.

Tracking and tracing of products and their ingredients is a common regulation, and the first section discusses some of the new regulatory requirements. IT discusses some new IT systems and technologies that you can use in track-and-trace systems.

Energy is an important element of manufacturing, and new regulations now require reports on energy consumption for production. This often requires collecting information from production equipment, HVAC equipment, and power equipment into a complete production report. The second section defines how information from these systems can be collected for regulatory reports.

Compliance with regulations requires collecting and safely storing compliance data. These data must be safely and securely stored, but there is increasing use in the IT industry to store data "in the cloud." The third section of this chapter discusses the issues you should consider when keeping compliance data off-site and "in the cloud."

Well-defined data retention and deletion policies are critical to regulatory compliance, even for manufacturing data. The last section in this chapter covers the issues and importance of deleting information and what policies you need to put into place to maintain compliance.

TRACKING AND TRACKING ON THE FACTORY FLOOR

New requirements for Manufacturing IT usually come from new business processes or new manufacturing processes, but sometimes they come from new laws and regulations. Two new laws, the USA Bioterrorism Act "USA Public Law 107-188" (www.fda.gov/EmergencyPreparedness/Counterterrorism/BioterrorismAct/default.htm) and the European Health and Consumer Protection Directorate 178/2002 (eur-lex.europa.eu/pri/en/oj/dat/2002/l_031/l_03120020201en00010024.pdf), will require new or expanded manufacturing IT systems. These regulations mandate the ability to rapidly trace the origins of all foods and all food additives in order to combat risks to the public's health and safety.

The basic concept of the regulation is that it should be possible to identify every ingredient in any food, trace the ingredients back to their sources, and the tracing information must be readily available. This requires that food producers maintain records that identify the ingredients for each product and the sources for each ingredient that goes into a product.

The Bioterrorism Act requires traceability known as "one up and one down." This means that food producers must keep track of where they obtained their raw materials and where they shipped their product. This is complicated because of the splitting and combining of material lots that normally occurs during food production. For example, the end product in a box on the store shelf will probably have had several different raw material lots from several different sources.

Usually, recipes and production schedules do not specify which raw material lots to use and it is at the discretion of operators to pick the appropriate material lot for each batch. The common practice today is to record this information manually, if it is recorded at all. Early FDA guidelines say that, in case of a problem with the product, the FDA can request exactly what was in a product, where the raw materials came from, and the manufacturer must reply within four hours of the request. The EU directive says that the information must be made available to competent authorities "on demand." The timing is critical; in the case of food contamination it can mean the difference between a safe recall and a disaster. Unfortunately, the four-hour and on-demand requirements cannot be cost-effectively met with paper-based tracking systems.

Fortunately, product traceability is a mainstream manufacturing requirement and is used in multiple industries. Material traceability has always been a regulatory requirement in the pharmaceutical and biotech industries and has recently been a requirement in consumer products due to product litigation concerns. These industries have implemented Manufacturing Execution Systems (MES) to automatically collect and manage tracing information.

However, MES systems are often expensive, complex systems, and hard to justify in many situations, especially if the only requirement is traceability.

This situation is changing because of the reference model for MES, defined in the ANSI/ISA 95.03 Manufacturing Operations Management Standard (www.isa.org). It provides a clear definition of the requirements for product traceability and most control vendors are now using it to model their MES solutions. Systems following the model should allow implementation of just production tracking, without requiring all of the other MES functions. Also production tracking can often be accomplished through the ERP system.

The technology required for production tracking is not exotic or advanced. It only requires a database and bar-code input from shop floor operators to identify material lots as they are added to products. Expensive solutions are not needed, but appropriate information technology must be applied to meet on-demand information requirements.

The changes coming to manufacturing IT include multiple new regulations, RFID, and wireless technologies. All of these changes will require more information technology to be applied on the shop floor and more integration of business systems and manufacturing systems. The modern control engineer must become familiar with basic information technologies and be ready to apply them to the new challenges in manufacturing.

WHAT'S YOUR ENERGY LEVEL?

There is another regulation that affects manufacturing companies. In July 2007 the European Union Directive 2005/32/EC for Energy-using Products (EuP) became effective. If you manufacture or design energy-using products and sell them in the European Union, then this directive has serious implications to both Business IT and Manufacturing IT organizations.

EuP is an eco-design directive aimed at reducing total energy usage and greenhouse gas emissions by any product that uses energy during its operation. It covers the full product life cycle, including the initial design and production of the product. The important element for Manufacturing IT is the inclusion of the energy consumed during material production and manufacturing.

Even if your company does not manufacture an EuP, it may produce materials used in an EuP and is, therefore, covered by the directive. Basically, Directive 2005/32/EC covers any product with a power plug, battery, fuel tank, gas, or steam hookup. The directive states that documents related to conformity assessments are to be available for inspection for 10 years after manufacturing and that the documents must be made available within 10 days of a request. This means that the record of the energy used during material production and manufacturing will usually be stored in an ERP or other business level system. Yet the source of the information in the documents will come from production-level systems.

Following the EuP directive will require business process changes, most of which will be minor, such as an inclusion of an energy element in the Bill of Materials (BOM). Business IT organizations may have already started projects to support EuP or they will be starting them soon. However, even minor business process changes may have a major impact on the shop floor systems that must collect and consolidate energy use.

Plan for EuP

Manufacturing IT groups should immediately start planning for support of the EuP directives. This includes determining where energy use is currently collected, where it can be calculated from machine run times, where it can be estimated from total energy use measurements, and where new sensors and measurements are needed. This information will need to be normalized to the same unit of measure, sanitized to remove extraneous energy usage (such as energy used during non-product-related maintenance or periods of nonproduction), related to the product being produced, and finally sent to the ERP system as part of a normal end of production report.

Initially, energy use may be estimated, but there will be growing financial and market pressures to report actual energy use. Actual use will become increasingly important over time as manufacturing efficiencies improve and these improvements are to be recorded as part of EuP compliance.

Integrate power, control, and HVAC

One unintended consequence of this directive will be the further integration of power distribution, HVAC, and production systems. Today these are typically separate systems, from different vendors, and managed by different departments. With the requirement to collect energy usage information (including HVAC energy use) as it relates to specific production, these systems will require integration. Manufacturing projects to support EuP will need to look beyond the typical production systems to provide a single consolidated source of energy use information. This may include integrating power distribution and HVAC systems into a plant-wide Manufacturing Integration Bus.

The Manufacturing IT skill set now includes knowledge of manufacturing-related laws, regulations, and compliance directives. The EuP directive joins the list of EU directives for manufacturing companies, including the Waste Electrical and Electronic Equipment (WEEE) directive, the Restrictions on the use of Certain Hazardous Substances in Electrical and Electronic Equipment (RoHS) directive, and the European Health and Consumer Protection Directorate 178/2002. See ec.europa.eu/enterprise/policies/sustainable-business/ecodesign/index_en.htm for more information on the EuP directive.

DON'T CLOUD YOUR COMPLIANCE DATA

CIOs are under continuous pressure to reduce capital assets, headcounts, and support costs, and cloud systems give them a way to meet those goals. Cloud systems are collections of IT resources (servers, databases, and applications), which are available on an on-demand basis, provided by a service company, available through the Internet, and provide resource pooling among multiple users. CIOs are looking for any and all opportunities to move internal company systems to external cloud systems because cloud systems reduce capital assets, IT maintenance costs, and direct labor costs. Companies are using cloud systems for e-mail, payroll, invoicing, ordering, and database management. There are over 50 different cloud service providers including powerhouses such as Amazon, Google, Microsoft, and IBM.

Cloud computing's silver lining is the direct cost savings. However, there is a dark center in the cloud when it is applied to manufacturing operations. In manufacturing companies, the operations and automation systems can be a major part of the IT budget, so CIOs will be looking to apply cloud computing to manufacturing and automation systems. While mission-critical applications with real-time response are not suitable applications of cloud systems, manufacturing databases may be a candidate for moving to the cloud.

Manufacturing databases can be huge, with hundreds of terabytes of data that must be safely maintained for years. This may seem like a good candidate for a cloud computing system, but the dark cloud behind the silver lining of cloud computing is that most external cloud computing services do not meet compliance rules. Manufacturing and automation operate under a large number of regulations and regulatory authorities, including OSHA, USDA, FDA, and EPA. All of the regulations have requirements of data integrity, security, backup, and nonrepudiation of data. This is difficult to achieve in cloud computing systems because you may not know where your data are actually located, what path it takes to get there, who really has access to the data, and how physically secure it is.

If your company is investigating a cloud computing solution that will include manufacturing data, then manufacturing IT must be part of the evaluation team. The manufacturing IT team must include personnel familiar with the manufacturing compliance regulations that affect your industry.

The assessment strategy should include checks for data segregation (which separates your data from other companies' data), a review of the provider's data leak prevention (DLP) process used to prevent insider attacks, the cryptography method used for transferring data across the Internet and storing it on disks, and the identity management process. Two additional important criteria to evaluate are the accessibility of the system for physical audits, which may be required by compliance authorities, and the availability of appropriate documentation that will be needed during an audit.

Consider using the Statement on Auditing Standards No. 70 (SAS 70) standards (sas70.com) in your evaluation. These define the standards an auditor would employ in order to assess the contracted internal controls of a service organization. Most cloud providers will have SAS 70 documentation. This does not tell you if they are secure but does show what security controls are in place. The Cloud Security Alliance (www.cloudsecurityalliance.org) also defines a set of best practices for providing security assurance in cloud computing systems.

Even the best security and data integrity processes and procedures may not be enough for your compliance data. Make sure that your company's legal advisors are involved in any "clouding" of compliance data to ensure that the risks of a compliance problem are balanced against the cost savings of moving the data into the cloud.

RETENTION AND DISPOSITION POLICIES—LETTING GO IS HARD TO DO!

Much can be said of the advantages of creating a manufacturing operations cloud, a set of servers that provide common services for manufacturing operations. These services include MES, LIMS, Maintenance Management, Warehouse Management, receiving, material tracking, equipment logs, and training logs. A manufacturing cloud can provide backup, archiving, redundancy, and scalability for these services. A manufacturing cloud can also provide the vital, but usually overlooked, function of deleting data.

There are many good reasons to keep manufacturing data. As engineers we want to keep data for comparison and evaluation because past production data can be used for lean manufacturing and Six Sigma projects. There are also rules and regulations that require data to be preserved. Data associated with materials, production, training, and maintenance often have compliance retention periods required by state and national regulations. Another important reason for retaining data is for legal challenges to demonstrate production quality and testing activities.

Despite the desire to retain data, there is an important reason to delete it. While deleting old and no longer relevant data can save disk storage and reduce backup requirements, the primary reason to delete data is litigation discovery cost. During the discovery phase of litigation, someone has to look at the data and decide if it is relevant to the case, and the time and cost to do the evaluation is large, often one of the major costs of litigation. Therefore, deleting old and no longer relevant data is a key requirement to reducing litigation costs. Companies that have been through a discovery process have realized that they need a formal Retention and Disposition Policy. This is a clear policy identifying types of retained data, how long it must be retained, and who has the responsibility for deleting it. Development of a retention and disposition policy must involve IT, legal experts, compliance officers, and line of business managers. It is important not to forget the people who create and use the content, because they may have local copies.

The retention and disposition policy and automated procedures must also have provisions for a litigation hold. In this situation, data that may be scheduled for deletion must be preserved if it is relevant to the litigation. This means that responsibilities must be assigned to ensure that relevant data are not accidentally deleted, because this is a red flag to juries that something is being hidden.

If your company has a retention and disposition policy, check to make sure that manufacturing data are addressed in the policy. Often the focus is on financial data and e-mails, but manufacturing data are often relevant to product liability litigation.

Also it is important to ensure that manufacturing implements the policies. If the company has a policy but does not follow it and ends up in litigation, then this indicates that the company cannot be trusted to follow any policies.

If your company does not have a retention and disposition policy, then start the work to support future policy development. This involves identifying all of the data that are retained, what the data represent, where data are located, where copies can reside, what applications are needed to delete the data, and who should have deletion responsibility. During development of the retention and deployment policy, you will need to know what the data represent in order to know what rules and regulations apply. These steps also help identify what data should reside in your manufacturing operations cloud. Permanently and unequivocally deleting data is hard to do, but it is an important function for well-run manufacturing IT organizations.

Building for Efficiency

One way to build efficient systems and cut implementation costs is through the use of well-established standards. The first section in this chapter defines some of the industry and international standards that manufacturing companies have been applying in order to reduce the time to develop production systems and also save money on implementations.

Efficient networks are required for production systems. They can place heavy demands on performance and consistency that are at odds with normal corporate network requirements. This chapter defines a practical network architecture that includes networks layers, firewalls, and applications, and describes how it fits into a corporate company-wide network.

Manufacturing Intelligence is the manufacturing equivalent of Business Intelligence (BI), and this chapter provides the background for understanding BI. It introduces you to some of the terminology of Business Intelligence and discusses the type of information used in Manufacturing Intelligence.

Efficiency in manufacturing often requires effective planning and scheduling, and there are many levels of scheduling in most companies. This chapter describes the levels of scheduling and shows you where manufacturing schedules fit into your company's planning and scheduling hierarchy.

The last section in this chapter is Service Oriented Architecture concept, and how it applies to manufacturing IT systems. A well-designed Service Oriented Architecture is a critical element of an efficient production system, and this section defines some of the rules and methods you can use to architect effective systems that will fit into your company's overall Service Oriented Architecture implementation.

CUTTING COSTS WITH MANUFACTURING IT STANDARDS

As every manufacturing company looks for ways to cut costs and save money, there is one place that too many engineers and executives are ignoring. If you had the opportunity to reduce automation project costs or times by over 30%, reduce costs for plant-to-enterprise integration by over 70%, or reduce maintenance support costs by over 10%, you would think that most engineers or executives would jump at the opportunity. Yet many professionals ignore these savings because they either don't know how to achieve them or they don't believe in the value of manufacturing IT standards.

The manufacturing IT standards

There are a set of well-accepted manufacturing IT standards that have a proven track record of reducing costs and time on manufacturing projects. The standards have been developed by multiple organizations including Mimosa (www.mimosa.org), OMAC (www.omac.org), ISA (www.isa.org), MESA (www.mesa.org), OPC (www.opcfoundation.org), and OAGIS (www.oagi.org/dnn2). Papers and presentations from these organizations document the savings.

These organizations are also working together through the OpenO&M initiative (www.openoandm.org). OpenO&M's goal is to provide a harmonized set of standards for the exchange of operations and maintenance information. It provides a forum for the committee chairmen to talk, coordinate their work, and remove duplications in the standards. Secondary goals are to develop a consistent implementation framework for the standards and a set of certification programs for the standards.

ISA standards

The well-known ISA 88 and associated MESA BatchML standards have a track record for reducing automation project costs by over 30% through code reuse and standard implementations, yet an estimated 50% of companies do not use these standards in new control projects.

Implementations of the reuse models defined in the ISA 88 Part 3 General and Site Recipe standard have reduced recipe development times by over 50%.

The MESA B2MML and ISA 95 standards have been used to reduce plant-to-enterprise integration project time and cost by over 70%. The ISA 95 standards for Manufacturing Operations Management (MOM) and Manufacturing Execution Systems (MES) have reduced analysis, design, and implementation of MES projects by over 30%, allowing more companies to obtain the well-documented productivity gains from MOM and MES systems.

OPC

The OPC and new OPC-UA standards for industrial device interfaces have reduced implementation and start-up costs through simplified diagnostics and management services. The OPC-UA standard defines standard interfaces to complex devices, such as analyzers, as well as provides an interface to legacy OPC devices. In addition, the OPC-UA standard addresses network security for industrial networks, eliminating the concerns that have kept OPC out of some facilities. OPC-UA is now also available as the IEC 62541 set of standards.

Mimosa

Companies implementing the Mimosa standards for maintenance information exchange have reported savings in maintenance operations costs. Just implementing the equipment identity section of the Mimosa standard reduces the confusion and lost productivity associated with multiple names for the same equipment in different systems.

Compliance testing cuts costs

The final part of cutting costs and saving money is to let someone else do the testing of systems. For the OPC standards, the OPC Foundation has started a compliance testing service. The service tests vendor packages against the industrial standards, thus reducing a company's need to test compliance and project integration costs.

Reuse to save

When looking to save money and reduce costs, it is important to reuse the work done by others. The Manufacturing IT standards, developed by industry experts, define the best practices for using IT concepts in manufacturing operations. They provide an excellent framework to help you do more with less.

NETWORK ARCHITECTURES FOR MANUFACTURING

As manufacturing operations, defined in the ISA 95 Enterprise/Control System Integration standard, move more into the standard information technology area, there are implications for IT organizations and network infrastructure layouts. IT organizations can gain value in faster and more robust implementations if they understand the real requirements for manufacturing applications and how this affects their infrastructure systems, such as networks and servers. This is required so that production can continue to have the robust and resilient systems they have come to expect.

ISA 95 levels

The ISA 95 Enterprise/Control System Integration standard, shown in Table 4, defines a multilevel model for activities. Each level provides specialized functions and has characteristics response times.

- Level 0 defines the actual physical production process.
- Level 1 defines the activities involved in sensing the production process and manipulating the production process.
- Level 2 defines the activities of monitoring, supervisory control, and automated control of the production process. It deals with time frames in the order of hours, minutes, seconds, and subseconds.
- Level 3 defines the activities of work flow, stepping the process through states to produce the desired end products. It deals with maintaining records and optimizing the production process. Level 3 deals with time frames of days, shifts, hours, minutes, and seconds.
- Level 4 defines the activities of establishing the basic plant schedule—production, material use, delivery, and shipping. It deals with determining inventory levels and making sure that materials are delivered on time to the right place for production. Level 4 deals with time frames of months, weeks, days, and shifts.

Table 4. The ISA 95 Enterprise/Control System Integration Standard

Level	Function	Description
Level 4	Business planning & logistics Plant production scheduling, operational management, etc.	4 – Establishing the basic plant schedule—production, material use, delivery, and shipping. Determining inventory levels **Time frame** Months, weeks, days, shifts
Level 3	Manufacturing operations management Dispatching production, detailed production scheduling, reliability assurance, ...	3 – Work flow/recipe control to produce the desired end products. Maintaining records and optimizing the production process **Time frame** Shifts, hours, minutes, seconds
Level 2	Manufacturing control Basic control, supervisory control, process sensing, process manipulation,...	2 – Monitoring, supervisory control and automated control of the production process
Level 1		1 – Sensing the production process, manipulating the production process
Level 0	Production process	0 – The physical production process

IT view of ISA 95 levels

While the models define a way to describe the activities of a manufacturing company they are often also used to define levels for software applications and networks. ERP, SCM, CRM, and PLM are typical Level 4 applications. MES, LIMS, CMM, batch management, and Warehouse Control Systems (WCS) are typical Level 3 applications. PLCs, DCSs, SCADA, and HMI are typical Level 2 systems and applications. There is an overlap, identified in Table 5, in the application area, with ERP often supporting some Level 3 functions, and DCS systems often supporting Level 3 and Level 2 functions.

These systems were generally never designed to operate together. Each level typically runs on a separate network with separate servers and under separate account and security domains, as shown in Figure 5.

As companies now work to improve their efficiency and productivity, these systems must be modified or designed to interoperate. They must coexist under the same IT infrastructure. Fortunately, there is a commonly used IT infrastructure that can be extended to connect the various application levels into a corporate infrastructure. This structure is based on the design of large-scale LANs and has been effectively applied in many companies and university systems, where there are thousands of connected computers across large sites.

The ISA 95 Hierarchy model can be applied to the equipment, people, and information systems in a facility. The model defines levels that can be

Table 5. System Overlaps

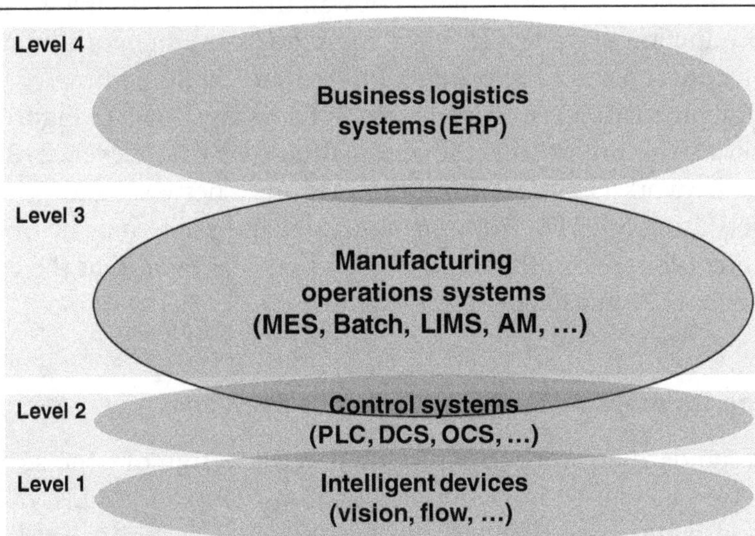

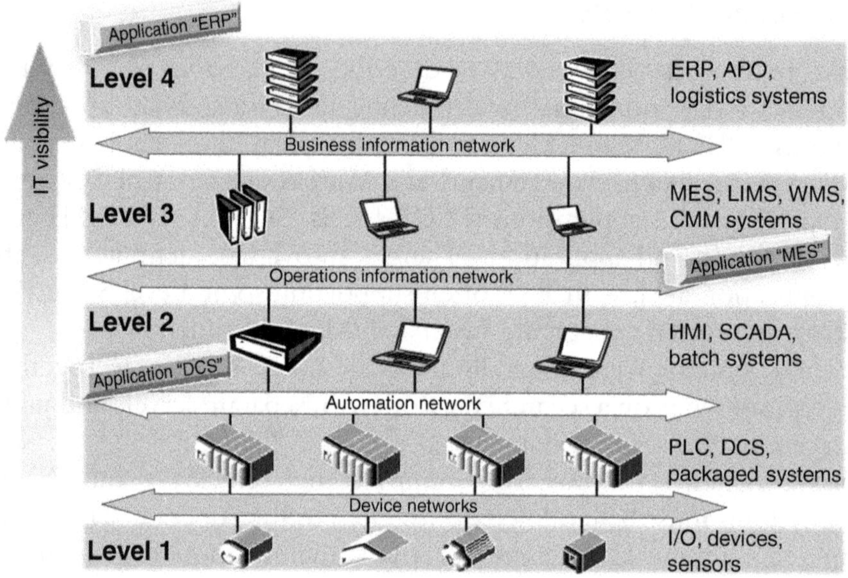

Figure 5. System networks.

mapped to systems and to networks. However, these systems and networks need to work together in a SAFE and SECURE manner. The ISA 99 Manufacturing and Control System Security standard expands on the concepts and recommends a DMZ between the Level 3 and Level 4 systems.

Virus and worm attacks to corporate systems should not be able to stop production systems, yet these systems can often not be as well protected and updated as business systems. You do not want hackers to be able to access and control production systems. To meet these goals the network infrastructure becomes a critical part of the production "system" architecture.

A typical physical view of systems and networks is shown in Figure 6 below. The physical structure isolates the automation network (Levels 2–3) and the operations network (Level 3) from the corporate network and other corporate subnets using a DMZ. Additional security and safety may be obtained by adding a firewall between the Level 2 and Level 3 networks, at the expense of additional network management tasks.

A structure for manufacturing applications networks

There is a standard network structure used in the design of large-scale LANS. This structure is defined in the book *Designing Large-Scale LANS* by Kevin Dooley and is also described in the Cisco Net 7 architecture definitions.

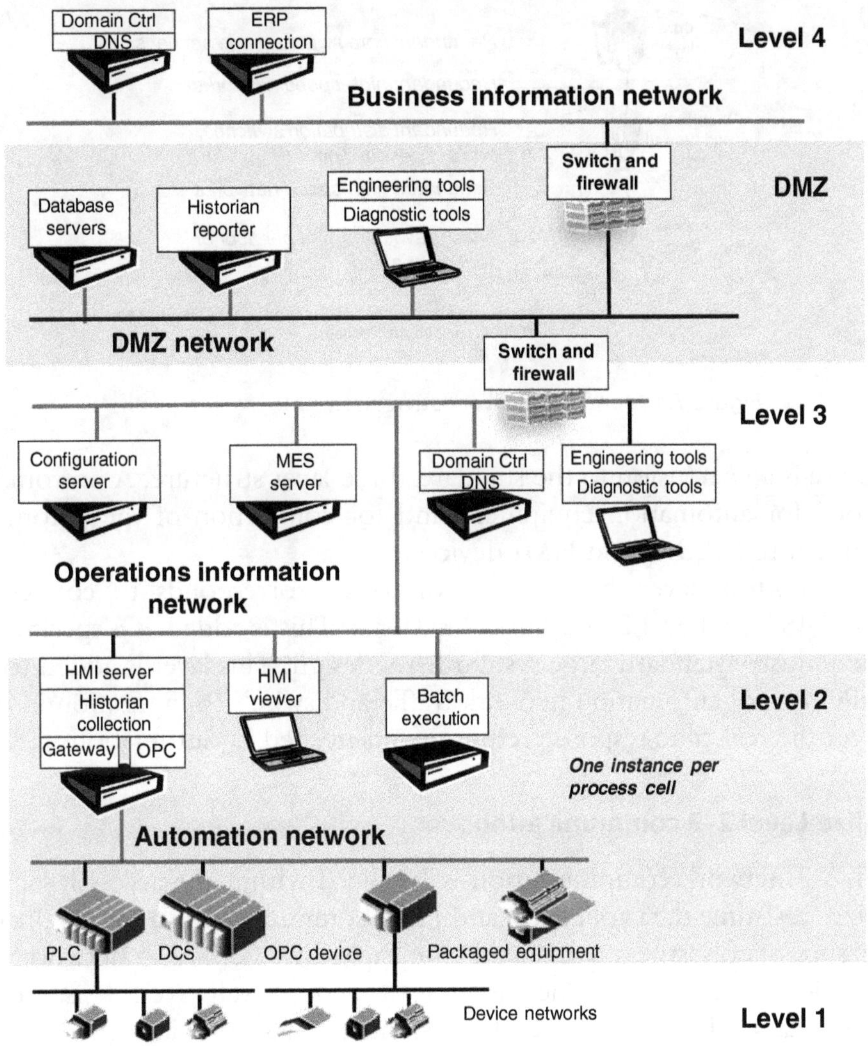

Figure 6. Physical view of systems and networks.

Many large-scale campus-wide LANs follow this structure, as illustrated in Figure 7.

Four-layer network model for control

An IT network infrastructure that follows the ISA 95/99 level approach is a four-layer physical network. The layers should not be confused with the ISA 95 levels, but they are instead layers of communication access, designed to maximize network uptime and minimize unnecessary network traffic. This

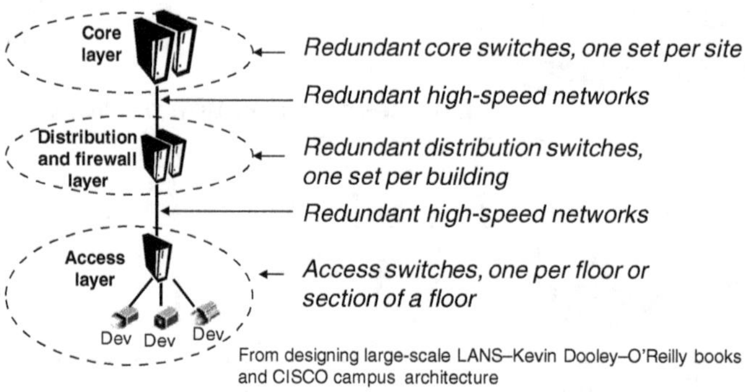

Figure 7. Standard network structure.

structure is an extension to the standard three-layer structure. A bottom layer is added for automation equipment and for connection of the automation equipment to SCADA and HMI devices.

The bottom layer, the automation layer, corresponds to connections between ISA 95 Level 2 and Level 1 systems. This is added as a specific layer to the industry standard large system LAN design. This layer is added to specifically control automation network traffic and switch configuration in order to meet the real-time response requirements needed by automation.

Localize Level 2–3 communications

Level 2–3 network communication is localized within the access layer. This localizes real-time data collection and HMI communication and also provides a measure of robustness and standalone capability. Level 3–4 network communication is localized in the distribution and firewall layer. This provides communication between all Level 3 systems and also provides a single point for firewall protection of the operations network (see Figure 8).

The manufacturing LAN Architecture is composed of four layers: Automation, Access, Distribution, and Core.

- **Automation layer**

  Automation devices that use Ethernet as Level 1–2 communication networks should connect to an automation switch. Only one device on the automation layer should connect to the access layer, and any communication should be through an application (such as a SCADA or HMI application). Other automation devices should only connect to the automation layer. The automation layer is a physically distinct network, completely separate from the operations LAN with no hub or switch connectivity.

  The automation layer should use switches rather than hubs to optimize network performance, especially if the PLCs and other control devices use the

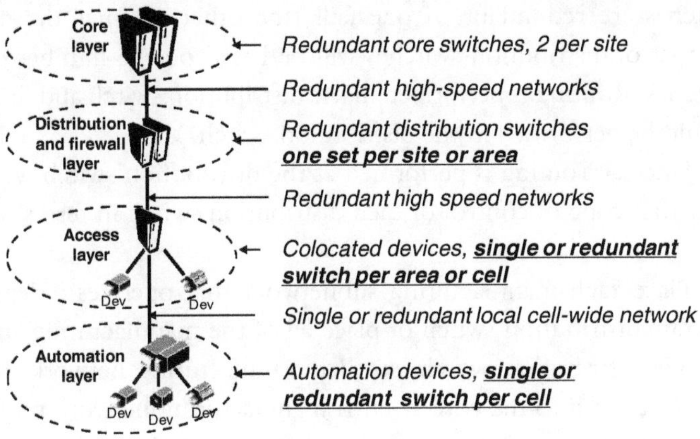

Figure 8. Network communication.

Ethernet network to intercommunicate. A hub logically connects all of the devices on to the same network, and all nodes see all communications. A switch will limit the traffic seen to each node, providing more consistent and reliable network response times.

- **Access layer**
  Other end devices (such as PCs, printers, and servers) should only have access to the Access Layer. To have redundancy, each Access Switch should be connected to two Distribution Switches. Consider using redundant access switches and two connections from each end device where complete failsafe operation is required. You should also consider the scope of control of each access switch in terms of the expected growth of the system, and the risk of failure of the access point.

  A standard corporate infrastructure for the business LAN will also use an access layer and have access switches. You should make sure that the business LAN access switches are kept physically distinct from the operations LAN access switches. Generally, the operations LAN access points cannot be modified at all without appropriate manufacturing and quality approval.

  The automation layer and access layer can be physically located on the same access switch; however, you should consider making each layer distinct by implementing a VLAN through the access switch. The devices on an automation LAN should all be on the same access switch to ensure that local automation network traffic does not need to be routed through the distribution layer.

- **Distribution and firewall layer**
  Distribution switches should be implemented in pairs. Each distribution switch in the pair has identical access switch connections. In this way the distribution

switches are redundant. If one fails, the other assumes the operation. The number of distribution switches will be based on the number of access point switches. Routing is performed in the distribution switch and firewall functions should be performed in the distribution switch.

Since all routing is performed at the distribution switch, you should consider the scope of control of each distribution switch in terms of the expected growth of the system.

Place each manufacturing subnetwork that operates independently on a separate distribution switch or place all of the manufacturing subnetworks on a single distribution switch, so that manufacturing network traffic does not have to cross into the core layer. If manufacturing network traffic crosses into the core network, then failures in other systems that affect the core network may impact manufacturing operations. If possible, let the distribution switch organization match the automation organization (process cell and area), since expected growth in most systems will impact a single process cell or area, or involve the addition of process cells or areas/buildings. Make sure that the business LAN distribution switches are kept physically distinct from the operations LAN distribution switches.

The firewall should be combined with the distribution switches, and generally high-end switches will include this functionality.

- **Core layer**
  The distribution switches connect to core switches. There are usually only two core switches on a campus network, provided for redundancy. The core switches consolidate the distribution switches. The sole function of the core switch is to transfer packets between distribution switches as quickly as possible with little or no overhead.

Company-wide networks

Figure 9 illustrates a typical campus or site-wide LAN. There are redundant core switches that only switch traffic between the distribution switches. The distribution switches are high-performance switches that route traffic between access switches and, in the case of a manufacturing sub-net, provide a firewall between the manufacturing and business LANs.

Local manufacturing nets

When the virtual separation is applied to the corporate network, then there are multiple virtual LANs (VLANs) created, each under a separate distribution switch, or a subnet under each switch, as shown in Figure 10. In this case, the

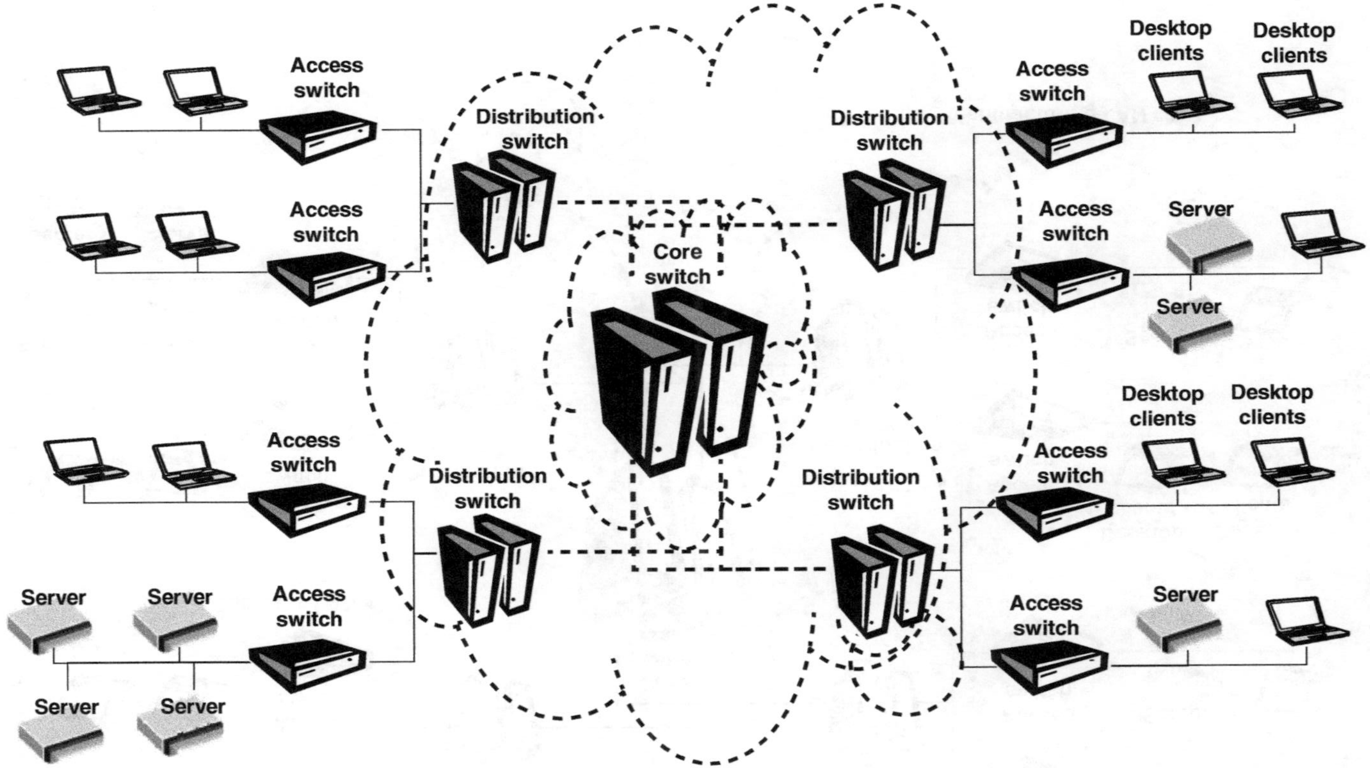

Figure 9. Typical campus or site-wide LAN.

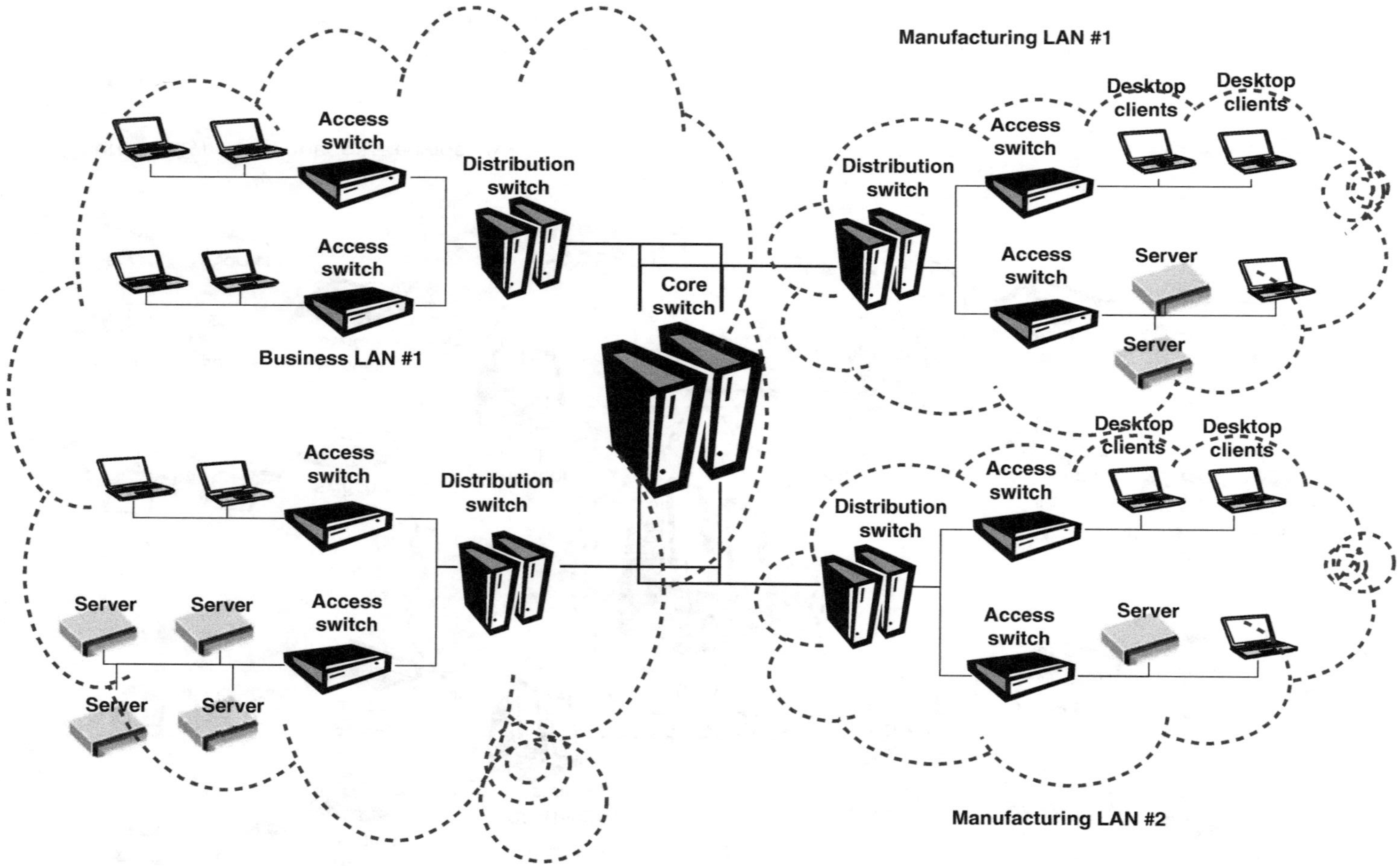

Figure 10. Multiple virtual LANs.

separate LANs should correspond to a scope of control of a manufacturing system, either a site or an area within a site.

Automation layer considerations

The left side of Figure 11 illustrates the two lowest levels of the network layout, with automation and access switches for a process cell or small area. The right side illustrates an alternate physical layout that uses a single switch used for the automation layer and access layer with the automation layer as a VLAN. While this provides a measure of protection, all automation and information traffic share a switch, and failure of the switch may stop critical control traffic.

The SCADA servers should control all communications to the automation layer through a gateway application. This controls the external network traffic allowed in the automation network, ensuring network responsiveness.

Firewalls

A manufacturing subnetwork should sit behind a firewall located in or at the distribution switch. Manufacturing applications are usually validated and cannot be easily changed. In some industries this involves significant testing and revalidation if any change is required. Due to this restriction, the manufacturing application servers and clients may not be running the latest anti-virus software. Some applications even preclude running antivirus software, since it interferes with the real-time processing required for the applications. Since you may not be able to execute virus protection on each system, access control to the systems should be strengthened to take up the load. Unprotected control systems are prime targets for infection, and they need multiple layers of protection. You can strengthen the first rule by adding firewalls between the control system networks and the rest of the corporate networks. This can be handled by a firewall at the distribution switch.

Firewalls with limited port access provide an important level of protection. The firewalls should be two way, in addition to protecting control systems from infection by corporate systems, it must protect corporate systems from the control systems. The purpose of the firewall is to protect both the corporate business systems and the Level 2–3 manufacturing operations systems.

Control at the access control routers can also be added to augment the firewall protection in the distribution switches. Access control routers allow only specified systems on the one side to access systems on the other side.

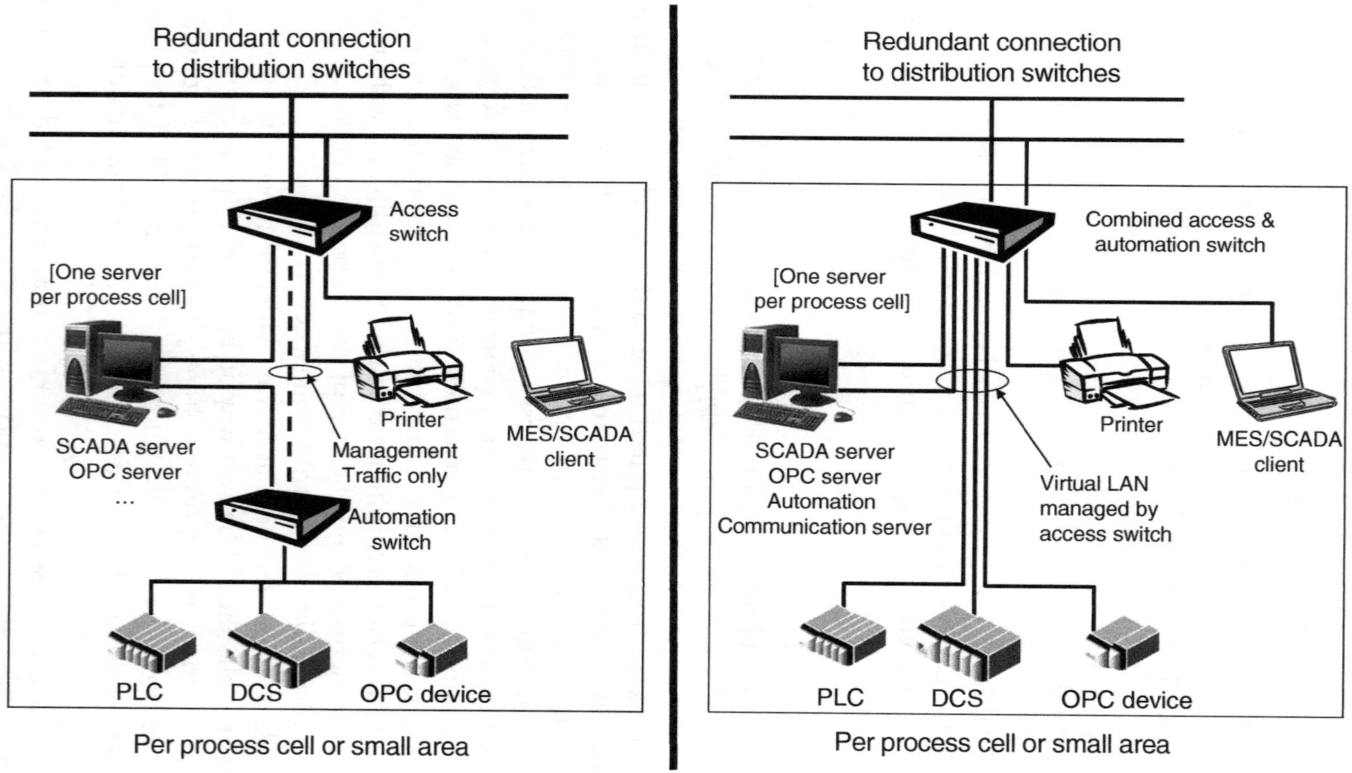

Figure 11. Two layouts for automation and access switches.

Summary

Applying the ISA 95/99 models to real manufacturing systems requires a detailed review and design of the supporting IT infrastructure. Fortunately, by extending the standard IT network infrastructure, it can meet the requirements for manufacturing applications. The extended infrastructure provides a separation of manufacturing LAN traffic from business LAN traffic, provides a location for a centralized firewall between business servers and clients and manufacturing servers and clients, and provides a special automation level for real-time systems with completely separated traffic.

This structure can usually integrate with the corporate IT infrastructure with minimal impact and disruptions. It allows for manufacturing applications and the network to continue operation in the event of loss to the corporate networks, either planned (because of business system shutdowns) or unplanned (because of network failures, virus attacks, or hacker attacks). IT organizations, which must integrate manufacturing applications and networks with business networks, can gain value by following the ISA 95/99 models in their infrastructure design. By applying the concepts of the models to a physical campus-wide or corporate-wide network they can design network systems that will be safe, secure, and reliable. In addition, the manufacturing systems can easily fit into the corporate infrastructure design and support organizations.

THE BEST LAID PLANS, OFTEN GO ASTRAY—WITHOUT CONSTRAINTS

Most manufacturing companies have a deep hierarchy of planning and scheduling processes that start with market forecasts and business plans and end with machine and unit schedules, as shown in Figure 12. The higher levels are the responsibility of logistics, marketing, sales, and executive management, while the lower levels of the hierarchy are typically the responsibility of manufacturing operations.

Manufacturing operations will become involved when actual production capacity is used in a finite capacity schedule.

A finite capacity schedule takes into account limited resources and determines a schedule that does not exceed the resource limitations. The limiting resources are often equipment, such as a maximum throughput. However, limiting resources could also be raw material availability, storage space, or personnel availability.

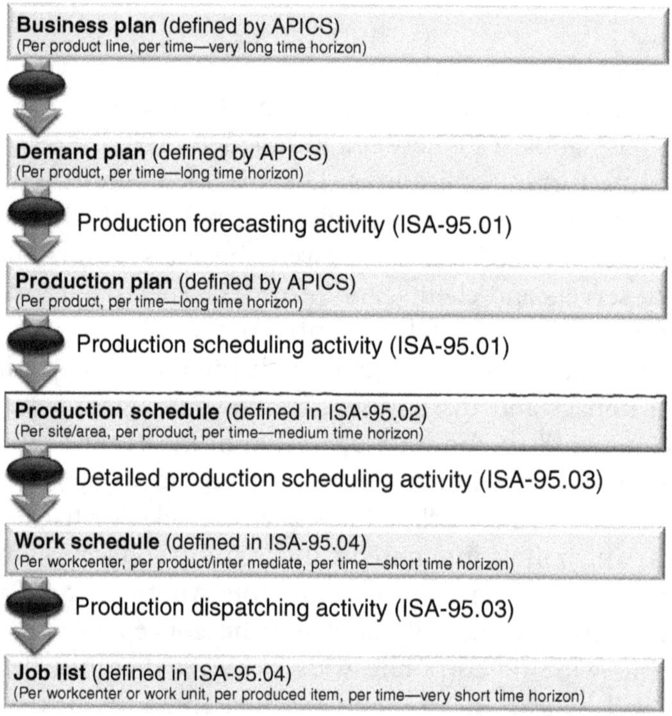

Figure 12. Manufacturing company information hierarchy.

Expose your capacity

Manufacturing IT's responsibility is to maintain and export the predicted production capacity so that it can be used to generate a finite capacity schedule. In many production facilities the capacity is constrained by a single resource for each production line, such as a bottleneck machine. These types of constraints are typical in discrete manufacturing and are usually product independent. In these situations the predicted capacity can be easily represented in a table of capacity per bottleneck for fixed time periods. The finite schedule time period, called the time bucket, can be hours, shifts, days, weeks, or months.

Consumable products may have time buckets of hours or shifts, while other goods usually have run time buckets of days and weeks. The time bucket will often be a compromise between production personnel, who want long periods of steady production for high efficiency, and supply chain planning personnel, who want small buckets for maximum flexibility.

Finite capacity scheduling systems often use Theory of Constraint (TOC) models and the Drum–Buffer–Rope (DBR) method for fixed bottleneck problems. These are explained in an easy to read series of books by Dr. Eliyahu Goldratt (www.goldratt.com), and should be required reading for any manufacturing IT professional.

Process flow scheduling

Process manufacturing often has "floating bottlenecks." This means that the bottleneck resource can change on the basis of current product or product mix.

Example:

- A single line may generate materials that flow into several downstream lines.

In these situations a more complicated scheduling method called Process Flow Scheduling (PFS) is usually used. PFS uses a model of the physical process and may be optimized for minimum material inventory or for economic manufacturing run lengths depending on the company's business needs. Representing the predicted capacity in a PFS scheduled system can be complex and complicated.

Because there is no single bottleneck machine, capacities must be maintained for each bottleneck machine in each part of a production line. Capacities must also be defined for different product mixes. This complexity usually requires a database or a set of tables, one table per product mix.

Committed, unattainable, and available capacity

In both discrete and process manufacturing the capacity information is also useful for operations management, providing a quick snapshot of committed capacity, the part of total capacity that is already committed to previously accepted production, unattainable capacity, the part of total capacity that is unavailable due to product mix, maintenance, or other reasons, and the available capacity that can be used for future production requests. Capacity information may also contain a confidence factor.

Example:

- The capacity information specifies what production will be available at 95% confidence and what additional capacity may be available at 50% confidence.

A confidence factor allows plant management to decide on the risk to take in accepting additional production requests.

Accurate capacity information is the goal

Providing accurate and timely capacity information to business scheduling systems should be a goal of every production facility. Plants need to maintain a database of capacity information so that they can receive accurate and implementable schedules.

SOA FOR MANUFACTURING

Service oriented architecture (SOA)

SOA is not a product or a technology or even an architecture. SOA is a set of principles for integrating software applications in a way that minimizes the complexity and number of interfaces between applications, therefore providing the most modularity for software evolution and support.

The principles of SOA define rules for deciding what software applications should support what functions and the type of information the applications need to exchange. As a result, the potential of SOA-structured systems is that they have clearly defined interfaces, that the interfaces relate to well-understood business or operational data, and that the applications are focused on well-understood business or operational processes.

Conversely, the downside of SOA is that there are still few SOA-related success stories because many IT organizations are using SOA tools, but are not yet implementing the SOA principles. SOA tools include Web services, enterprise service buses, XML gateways, simple object access protocol (SOAP) models, and representational state transfer (REST) models. These tools are the current best methods for application integration. However, just using SOA tools does not mean that systems follow SOA concepts.

SOA's role in manufacturing operations

The key to understanding what SOA means to manufacturing-related technologies and processes is to focus on software services. There are two general classes of services related to SOA: fine-grained services that deal with small pieces of data and coarse-grained services that deal with larger collections of data. The large collections of data are called business objects and the services are actions on the objects. Often a combination of coarse- and fine-grained classes is needed for any integration project, as depicted in Figure 13.

Coarse-grained services for actions

Coarse-grained services in the business world deal with objects such as purchase orders, invoices, checks, shipment receipts, transportation orders,

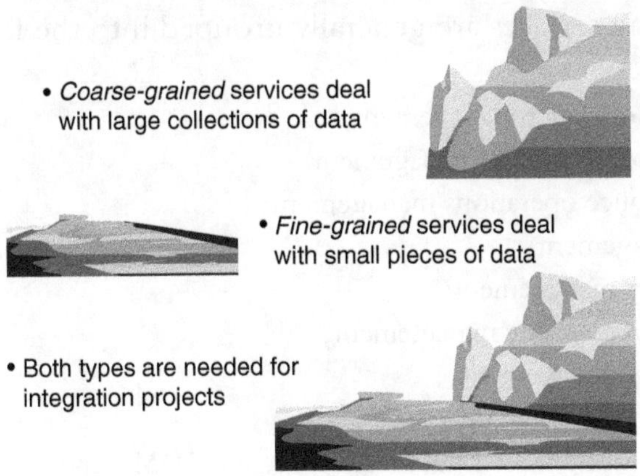

Figure 13. Illustration of SOA services.

advanced shipping notices, and inventory. Typical actions include creating purchase orders, sending advanced shipping notices, and moving inventory between plants. An SOA system would have integration interfaces that follow the action–object interface model (where the interface specifies the action followed by the object the action will be performed upon) with larger actions occurring on big and complex objects. Examples of business system interfaces in this instance would be things like: "create shipment receipt"; "make vendor payment"; "pay invoice"; and "create transportation order."

Coarse-grained services in the manufacturing world deal with objects such as production schedules, production reports, certificates of analysis, laboratory test results, maintenance schedules, and inventory. Typical actions include publishing of production schedules, creating laboratory test results, and moving inventory between production areas and warehouses. Examples of manufacturing system interfaces are "execute production schedule"; "move material lots"; and "assign material to storage container."

Fine-grained services for data

Fine-grained services are data query services, not action services. A query service deals with a small piece of data that is managed by one system but is also needed by other systems. Typical fine-grained queries include "get storage location of lot"; "get new lot ID"; and "get equipment status." If the system has only fine-grained action services and no coarse-grained services, then it is not SOA, even though it may use SOA tools. One common problem with SOA implementations is the use of a large number of fine-grained services, but with no partitioning of services by business objects.

Manufacturing services are generally grouped into the following areas:

- production order management;
- production response management;
- maintenance operations management;
- lab management;
- inventory management;
- operations capacity management;
- receiving management;
- shipping management;
- manufacturing master data management (MDM); and
- KPIs (key process indicators) and OEE (overall equipment effectiveness) calculation, monitoring, and management.

These services deal with organizations, locations, personnel, materials, equipment, containers, and tools. From an SOA perspective, the coarse-grained objects often relate to ANSI/ISA 95 objects such as production schedules, production performance reports, operations schedules, operations capabilities, and recipes.

Implementation models

There are two general models, shown in Figure 14, used for SOA implementations: an orchestrated model and a choreographed model. In the orchestrated model, there is a coordination service that contains the rules that sequence the coarse-grained services to implement the business process. Coarse-grained services in a well-defined SOA system will also use fine-grained services. The coordination services typically use business process execution language (BPEL) to implement business rules. BPEL is an OASIS (www.oasis-open.org) standard for specifying interactions with Web services.

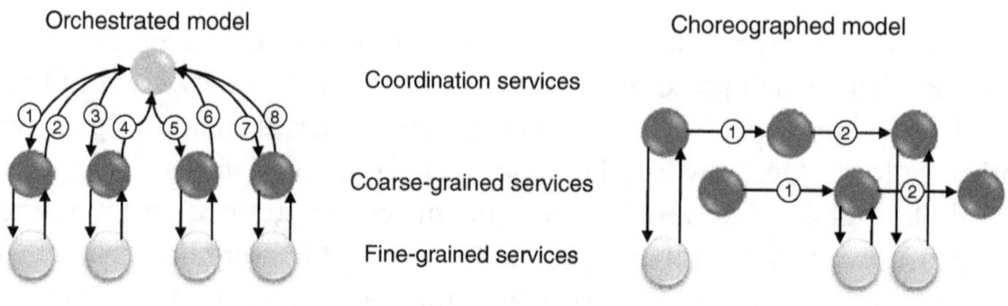

Figure 14. Two models of SOA integration.

Orchestrated integration

The orchestrated model isolates business processes in a coordination layer, allowing the coarse-grained services to focus on the business objects. Many modern business systems follow the orchestrated model because it improves modularity and reusability of services. Although there is a performance penalty using this centralized coordination service, the impact is usually not significant at the relatively slow speed of business processes. When real time is measured in minutes, extra messages and rule interpretation are significant only if hundreds of messages a minute must be processed.

Choreographed integration

Choreographed models differ from orchestrated models, in that there is no overall coordination layer in the choreographed model. A coarse-grained service may trigger downstream coarse-grained services. The business process is defined by the collection of rules implemented in the coarse-grained services. The choreographed model can handle faster process times and is therefore the model most often used in manufacturing applications and in customer-facing Internet applications. Both of these applications must handle thousands of messages per minute with minimal latency and delays.

Specifications and tools

SOA is based on six principles:

1. *Applications are loosely coupled.* Applications are loosely coupled when the availability of one system does not significantly affect the other system and when the implementation of a service is hidden from the requesting application.
2. *Interface transactions are stateless.* A stateless interface has no history associated with it; each use of the interface is based only on the exchanged data and does not use hidden knowledge maintained by the service provider.
3. *Interfaces follow the remote procedure call (RPC) model.* The RPC model means that the interface looks like a local function or subroutine call and the calling program does not have to handle any details of messages and communication protocols.
4. *Interfaces are message based.* A message-based interface sends messages between the applications using an enterprise service bus (ESB). An ESB provides fundamental services for complex architectures via an event-driven and standards-based messaging engine (the bus).
5. *Messages use XML data.* The messages are based on XML data—not flat files or a proprietary binary interface.

6. *Interfaces may support both synchronous and asynchronous transactions.* The services may be synchronous, that is, there is a request for the service and a wait for a response. Alternately, the interface may be asynchronous, with the application making a service request, while continuing to handle other processing, and the response arriving later.

SOA principles are highlighted in SOA specifications and SOA tools. The SOA specifications include the definition of loosely coupled business or operations processes, services that implement the processes, objects that contain the service information, and transactions that define the coordination of services and objects. SOA tools are applications, application services that implement business services, standard messages that represent the business objects, and integration services that implement the transactions.

SOA in manufacturing

SOA will affect manufacturing systems as their use becomes more ubiquitous in internal and externally linked applications. As a result, operations systems and control systems will have to operate in an SOA environment and coordinate with other SOA-defined applications.

This means that control applications will need to expose their core business functions as loosely coupled coarse-grained services and expose their shared data through fine-grained synchronous services.

For example, a coarse-grained service may be a command to an Automated Guided Vehicle (AGV) to move material between a warehouse location and a production location, while a fine-grained service may be a request for the bar code number of the pallet currently being moved.

Application of SOA principles will simplify manufacturing systems because the principles are consistent with the best practice models of loosely coupled business and manufacturing systems. SOA principles are also entirely consistent with the best practices for manufacturing system security. To keep your systems up to date with the increasing use of SOA, you should look to begin widely applying SOA principles in your manufacturing applications.

Personnel, Equipment, and Material

Three main physical entities are used in manufacturing: people, equipment, and material. One of the most important Manufacturing IT tasks is to keep track of these physical entities, and this chapter addresses information management issues you should be aware of with personnel, equipment, and material.

Managing equipment provides a challenge for manufacturing companies because of the many identifiers that can be associated with a single piece of equipment. Manufacturing equipment is used by many different departments, usually using many different vendor systems. This chapter identifies some of the different systems that you will use to identify and track equipment.

Manufacturing is all about materials and the transformation of materials. Information about materials must be shared between the business systems that purchase and sell materials and manufacturing that consumes and produces materials. This chapter discusses what information about materials must be shared, some common names for the shared information, and who "owns" the information.

Personnel are a vital aspect in manufacturing, but they must be trained and qualified to operate the production equipment. IT systems provide methods to identify people and associate them with roles and rights to perform actions. These systems are a vital part of any manufacturing system and should be part of your manufacturing solutions.

WHAT'S IN A NAME—EQUIPMENT?

The "Information" part of Manufacturing IT is as important as the technology part. One important element of managed information is the correct identification of the equipment used for manufacturing. Equipment identification is used for:

- scheduling,
- qualification,
- tracking and tracing,
- maintenance,
- troubleshooting,
- visualization (HMI),
- finance,
- capacity tracking,
- Overall Equipment Effectiveness (OEE), and
- multiple other reasons.

Unfortunately, it is not uncommon for a manufacturing company to have four or five or more different identifications for a single piece of equipment. Therefore, a critical aspect of equipment information management is managing different equipment IDs across multiple vendor systems and applications.

Physical asset naming

There are two main categories of equipment identification:

- an identification of the role that the equipment performs independent of which specific piece of equipment is performing the role.
- an identification of the actual physical equipment.

The ISA 95 Enterprise/Control System standard supports these categories by defining a role-based equipment model and a physical-asset equipment model. The models also show how the two ID categories are related over time as equipment is replaced during maintenance operations.

Even within these categories, there are different identifications. Each physical asset has a vendor serial number that is needed for financial tracking and maintenance contracts. Vendor serial numbers are not unique, so each physical asset is usually given a unique company-specified financial asset tag number. The financial asset tag number is maintained in a financial system and may not be compatible with maintenance system identifications.

Usually differences are in the number of characters or numbers used and the naming rules required by the applications. Therefore, maintenance systems often have a separate maintenance tag that is added to the physical asset. In the case of small equipment, the three separate identification tags and vendor make and model information can cover most of the equipment's surface area.

Role-based naming

Within the role-based category, there is an identification of the equipment as seen by a SCADA or HMI system. This usually corresponds to a common functional name or a name documented on a P&ID drawing. Individual pieces of equipment often follow the ISA naming conventions, while collected elements (such as units) follow company-specific naming conventions.

Data historians also require equipment identification. While the historian equipment tag names are often the same as the SCADA/HMI names, if systems from different vendors are combined then the IDs can be different. MES systems will often use the SCADA/HMI names for equipment, but name inconsistencies can also occur because vendors require different ID rules.

ERP logistics systems also identify equipment by role-based identifications for scheduling or planning. These names are rarely the same as the SCADA/HMI names and are often just resource ID numbers used for scheduling purposes.

Finally, there may be a custom qualification management system used to maintain the qualification status of equipment for specific products or processes. The scope of all of these IDs varies widely, from a local area scope for an HMI display to the entire company for the financial asset ID.

Use a standard to save your names

Managing all of these separate IDs and keeping them consistent as new equipment is added, obsolete equipment decommissioned, and equipment replaced can be a huge task, typically requiring a shared database of IDs or storage in a Master Data Management (MDM) system. However, equipment ID management is a vital aspect of implementing the tracking and tracing functionality required by federal and state regulations and must be part of every Manufacturing IT's scope of responsibility. Consider using the ISA 95 role-based equipment and physical-asset equipment models as the basis for describing and storing the IDs associated with your production equipment.

WE LIVE IN A MATERIAL WORLD

Manufacturing lives in a material world. It is not surprising that manufacturing systems have a requirement to identify and track materials because the main purpose of manufacturing is to convert materials in one form into materials of another form. An important part of manufacturing system integration with other corporate systems is maintaining and exchanging material identification and information. The source of the material identification usually resides in an

ERP. The ERP contains the material masters, the information that is used to identify types of materials and vendors, and the information used in inventory, accounting, ordering, receiving, and shipping.

However, material information is required in Manufacturing Operational Management (MOM) systems such as:

- MES,
- batch management systems,
- tracking and tracking systems (bar-code scanners and RFID readers),
- Laboratory Information Management Systems (LIMS),
- Warehouse Management Systems (WMS).

Shared material information between the ERP and MOM can be divided into three main categories:

- material class information,
- material definition information,
- material lot information.

Material classes

Material class information identifies materials regardless of the source of supply of the material. Material class identification is used in Bills of Material (BOMs), recipes, and assembly instructions.

Material classes contain non-vendor-specific nominal values for material properties, such as weight or color.

Examples:

- Material classes could be magnesium stearate, 2 inch #8 brass screws, and 50 mil aluminum sheeting.
- A BOM may specify that 200 kg of magnesium stearate is required but will not normally define the material by source?

Material definitions

Material definition information identifies material from specific vendors or sources. The material definition is used to identify material from qualified suppliers and to identify specific material sources for tracking, ordering, and generation of "equivalent-material" lists. Material definitions contain vendor-specific nominal values for material properties, such as percent purity or hardness.

Examples:

- Magnesium stearate from Alpha Chemical identified as MS-0023AC.
- Magnesium stearate from Beta Chemical identified as MS-0023BC.
- Locally produced magnesium stearate as MS-0023XX.

Material lots

Actual material is identified by material lot. Each material lot defines a specific quantity of the material and properties of the material. The properties may be determined locally through incoming inspection and laboratory analysis or may come externally through certificates of analysis. Material lots are used for tracking the status of the material, such as provisional, tested, released for production, or released for shipment.

There are some specific aspects of material lots:

- Material lot identification is required for tracking and tracing and the creation of a material genealogy for products.
- Material lots may be stored in multiple containers, such as a single material lot of screws that is stored in multiple cartons.
- Each separately stored amount is identified as a material sublot. Material sublots may have separately maintained quantity and location information, and the sublots may also have unique properties such as RFID identification or qualification status.

Sharing the material information

Material class, material definition, and material lot information are used in both ERP systems and MOM systems. Material sublots are used at the MOM level but may also be used in the ERP system, if the ERP system tracks inventory or schedules production below the lot level.

Usually the ERP system is the custodian of all material information, while the manufacturing systems update the information with quantity and location changes and with new values for material properties. A common practice is to keep copies of the ERP information at each manufacturing site so that the information is available when the ERP system is unavailable.

Many companies are now using the MESA B2MML standard (www.mesa.org) as the integration standard between their ERP system and manufacturing systems, and often the first integration project is sharing material information. When looking at sharing information across systems, remember to handle all three material information types, material class, material definition, and material lot, in your integration projects.

SOMETIMES IT'S PERSONNEL

Sometimes it's easy to forget about the people aspect of control systems. Multiple regulatory rules, laws, and internal procedures require that people who perform shop floor actions are unequivocally identified, are authorized to perform the actions, and have valid training or qualifications to perform the actions. Because personnel information is usually maintained in multiple IT and control systems, it is a key area of exchanged information.

Which system checks and confirms the identity of personnel in your environment? Is the master information used to identify a person in a:

- network security system,
- in an IT user management system,
- in an HR system,
- in a control system,
- or in all four systems?

Many organizations are moving to a Single Sign On (SSO) environment in which a user logs in once and all applications have access to the user's credentials through an SSO management application. The leading SSO system is based on the Liberty Alliance consortium (kantarainitiative.org) and uses the OASIS XML-based security services (www.oasis-open.org) called the Security Assertion Markup Language (SAML).

Keeping accounts up to date

Unfortunately, many control applications have their own user account management and require manual synchronization with the other systems. It is common, but not desirable, that control systems contain old accounts for people who have left the company or have moved to other positions.

A critical Manufacturing IT function is to keep the control system security up to date with changes in personnel information. Manufacturing IT teams should have a "tickler" system to remind them to make periodic updates or they may subscribe to RSS feeds from the HR system to be informed of pertinent personnel changes.

Identifying a person and determining if they are authorized to use a system to take a control action is often not enough. Regulations and company policies may also require that the person is qualified, or appropriately trained, to take the action.

Some MES systems provide a training and qualification tracking function, providing a real-time check of qualifications before a user is allowed to take an action. The qualification information may also be maintained in a corporate training tracking system or the HR system. The corporate systems normally do not have the real-time availability or up-to-date data required for just-in-time qualification checking; therefore, the MES and training tracking systems must share information and be synchronized.

Some corporate policies may also require that qualification certification information, such as state-issued professional license numbers, are included in production records. This is another type of information that is usually maintained in HR systems but is required in the real-time operational systems. There is a standard format for exchanging this information using the B2MML Personnel Schema standard (www.mesa.org), but few vendors currently support this functionality so most synchronization of qualification information is performed manually.

Where are you?

In addition to authorization and qualification to take an action, there is also the question of who is nearby to take an action. Location information may be managed by a door/room security system that tracks entry and exit or through GPS tracking of cell phones tied to a physical security system.

This information is extremely useful to assign actions, alarms, and events to operators based on their location. Personnel location information is also of vital importance in case of emergencies or hazardous conditions. The B2MML Personnel Schemas can also be used to exchange this information.

Often keeping personnel information in sync with business and control systems is a low-priority task, until inspectors or lawyers get involved. Make sure that the personnel information in your control systems is up to date and in sync with corporate systems to prevent any future personal problems with personnel information.

Knowledge

The last chapter deals with managing knowledge in your Manufacturing IT organization. Knowledge is complex and managing knowledge capture and transfer usually involves many different systems and vendors. However, managing knowledge gained during a project is important because it can be very transitory, quickly lost, and difficult to reacquire.

This chapter introduces you to some of the issues associated with knowledge management and some of the technologies used to acquire and maintain information. One of these technologies, specifically the WIKI technology, can be effectively applied in all areas of manufacturing. The first section in this chapter discusses using WIKIs to help solve knowledge capture problems.

It is also important to know the standard policies and procedures that can be used to manage software systems, because these policies and procedures can be effectively applied in manufacturing IT systems. The second section in this chapter defines the Information Technology Infrastructure Library (ITIL), which are best practices that should be applied to IT system maintenance.

Finally, it is important to know when and how to fold an IT project. The last section in this chapter discusses how you should capture knowledge on a project, even an unsuccessful project, so that later projects will be efficient and effective.

QUICK, QUICK—KNOWLEDGE MANAGEMENT

It is hard to believe in the information age that one of our biggest problems may be the loss of vital information.

Three factors are converging:

1. fewer manufacturing professionals,
2. geographical distributed work environments,
3. a large percentage of manufacturing professionals nearing retirement.

Information and, more important, knowledge, which has been patiently collected through years of experience, will be lost because engineers do not have the time or opportunity to pass on their knowledge to the next generation of engineers. Information loss is especially critical in the area of control systems. These systems often have lifetimes measured in decades, and when the time comes to upgrade or replace them, the engineers are often retired or have left the company.

Capture knowledge before it flees

Knowledge has been captured on paper; however, paper is expensive to maintain, difficult to share, and slow to search. The IT industry has been working on the problem of capturing knowledge and has developed solutions called Knowledge Management Systems. These are tools that capture, organize, and store knowledge of individuals and enable easy sharing of the information. Knowledge management systems are also used to formally record and distribute information that has been learned through experiences and studies.

However, Knowledge Management Systems suffer from flaws for use in most control engineering. They are usually implemented in companywide projects with large budgets, large staff, and an equally large effort to capture company knowledge. They provide a structure for the captured knowledge, and sometimes require forcing knowledge into predefined structures.

Use a Wiki

This is the case where the old can learn from the new. As students know, one of the main knowledge references on the web is Wikipedia (http://www.wikipedia.org). Wikipedia is an on-line encyclopedia that is built through public collaboration. Anyone can add to it or edit existing information. It is the ultimate open repository of public knowledge, collectively managed and maintained by the public at large. Anyone can edit any article, so misleading, out of date, or incorrect information can be entered. However, because experts monitor contributions and there are many readers, incorrect information is usually quickly found and corrected. Students use Wikipedia to rapidly find otherwise obscure pieces of knowledge, which usually have links to more detailed sources and references. A key point is that Wikipedia is a knowledge management tool for unstructured knowledge, and it is based on technologies that can be applied to control system knowledge.

Quick, quick

Wikipedia is based on a Wiki web system. Wiki means "quick" in Hawaiian, and Wiki web sites are often called wikiwikis (or quick-quick). The first Wiki was created in 1994 for collaboratively publishing software patterns on the web. Wiki web sites have a simple editing process, allowing nonprogrammers to quickly and easily add information and links. A typical Wiki is made up of thousands of linked pages and search methods. Authors can set up tracking so they are informed when changes are made, providing a method for analysis and review. All changes are saved and previous versions of pages are available on-line.

Engineering groups can set up their own Wikis as inexpensive repositories for knowledge. Wikis do not require any predefined structure for knowledge; the structure grows as information is entered. Wikis provide an easy way for engineers to record experiences and events, and for others to find and extend that knowledge. The ease of editing the Wiki is one of its greatest assets: an easy-to-edit resource increases the compliance of engineers in maintaining the wiki as a storehouse of information.

Open source Wiki tools

There are many open source Wiki tools available, based on PHP, Java, and Python (web programming languages). See www.wikipedia.org/wiki/List_of_wiki_software for a list of Wiki choices and www.wikipedia.org/wiki/Comparison_of_wiki_software for a comparison. Keep in mind features that you'll want to have in your internal Wiki, such as ability to run on an intranet, access control, and the ability to write in Latex, if necessary. Don't forget that you can download multiple Wiki programs to test it as a single user, checking whether they are easy to use and meet all your requirements, then save your work as HTML file to use in your final choice.

A collaborative knowledge base

An internal Wiki can become a valuable tool for a collaborative knowledge base. If other information such as internal documents, vendor manuals, reports, and studies are available on-line, then they can be easily linked into the Wiki site. An internal Wiki can be a simple answer to the complex problem of control system knowledge management. Visit Wikipedia for more information and start preserving vital control system knowledge before it is lost.

DON'T DROWN IN MANUFACTURING PAPERWORK

Finding vital business information can be a time-consuming and frustrating task. Office personnel, vendors, and customers can generate volumes of paper. Yet the volume of office information can pale in comparison to the paper records generated by many manufacturing companies. Some companies estimate that knowledge workers (just about anyone using a computer on a regular basis) spend up to 20% of their time just trying to find the paper records they need for their daily tasks. In manufacturing companies the numbers can be worse with operations personnel locating records for engineers, scientists, and quality personnel. When manufacturing records are kept on paper it is common to take months to find vital records. This delay is so well known that the FDA gives companies 18 months to resolve problems and up to 80% of that time is used to find the right paper records.

Scan it, save it

There is an IT solution to this problem, and it's low-tech than many other IT solutions. The IT solution is to scan in every paper record and convert them into searchable PDF files. Other industries, such as health care, insurance, and legal, have had to deal with massive amounts of paper, and their requirements have led to the development of high-speed document scanners.

High-speed document scanners that can scan over 60 pages per minute with document feeders for hundreds of pages and standard interfaces are available from multiple sources. The scanners can be connected to an application server to automatically create a useful file name, set document properties, and save the scanned document into the proper folder. The files can be saved on simple file servers, document servers, or collaboration servers such as Microsoft SharePoint.

All of these storage methods should be integrated with a search engine, letting personnel quickly find the right data. The best part of this solution is that the operational cost can be low. A typical installation, using existing file or collaboration servers, can have an operational cost of less than $1 per scanned record, and scanned records can be large with batch records often over a hundred pages. Typical paper records that should be scanned include maintenance records, production records, laboratory records, and inventory record.

Look for wasted time

A quick way to see if this low-tech IT solution will help your bottom line is to perform a small intracompany survey. Identify those who are making the information requests and find out how much time they waste trying to obtain the paper-based data they need. Identify the people actually searching for the right paper records and find out how much time they spend searching. On-line access to that data can easily save 80% of the time, giving you a quick estimate of savings.

Standardize names and terminology

Once you have started the project to scan in all records, then the next step is to standardize names and terminology. You can start by developing and implementing standardized names for equipment tags, product quality attributes, process parameters, process data tags, operator identification, and operations or manufacturing method names. Standardized names will simplify searches and assist in cross-site investigations and studies.

Solving your paperwork problem

Companies that have implemented a standard scanning procedure and naming conventions have found that it is invaluable in Six Sigma and lean manufacturing projects, reducing the data collection time significantly. Remember, the on-line records do not replace the paper records. Often the paper records are needed for regulatory and legal reasons, but on-line records reduce the need to index, search, and otherwise use paper records. If you are still drowning in manufacturing paperwork, then it is time to look at document scanning to help solve your paperwork problem.

WHAT'S THE ONE RIGHT ANSWER, ENTERPRISE SEARCH FOR MANUFACTURING?

Today you can find volumes of information from any place in the world within seconds using Internet search engines. Students use this capability to find more information in minutes than previous generations could discover in weeks. As these students enter manufacturing companies, as employees, they bring an expectation of the same search capability for corporate information. They expect that all information should be quickly searchable and available.

Today you can find the menu, reviews, and a map to an unfamiliar restaurant within seconds, you can browse library archives on the other side of the world without leaving your desk, but in many companies you cannot find last month's production numbers or equipment logs in less than a day of searching.

Enterprise search is different

IT departments are addressing this information problem through Enterprise Search projects. Enterprise Search describes applying search technologies, such as Google and Yahoo, to corporate data. Enterprise search differs from Internet searches in several fundamental ways. Corporate data differs from publicly accessible Internet data, it has numerous security and confidentiality concerns, the data are in many formats and databases, and often special applications are required to read the data.

Internet search users are looking for any answer and expect to see dozens or hundreds of documents, enterprise search users are looking for the single right answer often in a single document. Internet information is usually hyperlinked, allowing users to trace through connections to find information, while corporate data are rarely hyperlinked. Internet data are frequently searched, so relevance can be determined based on access counts.

However, there are fewer enterprise searches, and relevance must be determined by content. Corporate data are often "dirty"; there are many misspellings, different terms for the same concept, the use of acronyms, and department-specific jargon. A significant fraction of corporate data is in e-mail, but there are privacy and confidentiality concerns with making e-mail searchable. All of these differences make it difficult to find the single "right" answer to an enterprise search.

Get involved in enterprise search

If your IT department does not yet have an Enterprise Search project, then they are probably investigating the technology. It is important for Manufacturing IT to be ready for enterprise search when it rolls out. Enterprise search is seen as a basic IT service, similar to databases and e-mail. The search service is managed by the IT department, but the content of the search is managed by the end user organizations. This is an important distinction, because it means that the content owners have some responsibility for enterprise search.

An IT department may have a "search manager" who is responsible for the corporate search servers and search application. You should identify manufacturing IT "content managers" who are responsible for an area, such as production or laboratory area. The content manager's responsibility is to make sure that any changes to information location, all new systems, and all deleted systems are communicated to the search manager. The content manager should determine what files, databases, and directories are to be included in the enterprise search.

Typically searched manufacturing documents are batch records, validation documents, maintenance records, qualification records, test results, material information documents, equipment use records, sensor logs, and calibration logs. The content manager should also determine how frequently the information should be re-indexed by the search engine. Re-indexing is a network-intensive task so it is a good idea to set up separate directories and databases for long-term storage and for current information.

Determine your content manager

Meeting the expectations for enterprise search will require new roles and responsibilities in Manufacturing IT, so it's important to be prepared for the coming wave of Enterprise Search projects.

"HOW MAY I HELP YOU?"

How many times have you answered your work phone, "Hello, how may I help you"? If you design and implement systems, you may be surprised at the amount of time spent on service and support. The typical IT department will spend over 75% of its budget on service. Control departments may spend an even larger percentage of their time on support of existing systems. The typical control department will be supporting PLC programs, SCADA configurations, HMI displays and reports, MES software, PC-based software, and a wide range of special applications and hardware systems. The control department may also have to support the application computers, file servers, print servers, their operating systems, the attached fieldbuses and industrial networks, network routers and switches, bar-code scanners, printers, and virus protection software. The control department may have to manage accounts and passwords for operators and imbedded software.

A contemporary control department often has the same demands for service and support as a contemporary IT department. IT departments that support financial, communications, and other vital services have developed "best practices" for handling service and support in a consistent manner, and control departments can learn from their experiences.

ITIL be some help

IT departments are increasingly implementing a best practice model developed by the UK government's Central Computer and Telecommunication Agency called the Information Technology Infrastructure Library (ITIL). ITIL is described in a set of seven books that are now controlled by the UK Office of Government Commerce. ITIL complies with the requirements for ISO 9001 quality standards and is referenced by ISO 9000 for IT standards. The ITIL models define a comprehensive process-based framework for IT service management. They contain a set of best practice guidelines for service support, service delivery, application management, infrastructure management, implementing service management in an organization, determining the business value of service, and implementing security requirements.

Focus on subareas

Service management and service delivery have subareas. Service management includes: configuration management of assets and their interrelationships, incident management and processes for getting service restored, problem management and process for determining the root causes of problems and correcting them, change management and the processes for ensuring that all changes to managed elements are carried out in a planned and authorized manner, release management and processes with managing updated to hardware and software, and service desk functions and process that describe how to management service requests.

Service delivery includes: dealing with end-user (customer) needs, ensuring high service availability, managing capacity to maintain performance, financial management for service costs, continuity management in the case of natural or man-made disasters, and organizing and maintaining the management of IT security. Many of these processes are also handled by control departments, but they are often handled in an ad hoc or nonsystematic method.

ITIL provides a solid framework for identifying the processes that should be in place for managing changes and problems in a computer-heavy control environment. The ITIL toolkit and books are available online for a small cost. See www.itil-officialsite.com for additional information.

Control departments will probably find the most value in the service support, service delivery, and security management books. Control department managers may find value in the business perspective and implementation planning books.

How do you ITIL?

ITIL defines what must be done for effective service, but not how to do it. For example, ITIL specifies that a service organization should allocate a priority to each incident but does not specify how to determine priorities. The service organization fills in the details of the process so it matches its problem space and organization structure.

ITIL complements other methodologies. The CMMI model addresses application development, the ITIL framework addresses what happens during and after installation of the developed software. Six Sigma and other improvement methodologies provide guidance in how to measure and improve software development and service delivery.

Apply ITIL to your control system support

Control departments should look at ITIL when designing their support service processes. The best practices in IT departments can be effectively applied to IT-intensive control systems.

KNOW WHEN TO HOLD 'EM, KNOW HOW TO FOLD 'EM

Many manufacturing IT projects do not finish, not necessarily because they failed, but because of changing requirements, changing business conditions, company mergers, or changes in products and processes. It's important to know when to fold a project, but it is equally important to know what to do when you fold a project.

A manufacturing IT project contains more than just software. Any project, if it a PLC program, DCS program, MES system, SCADA system, HMI system, integration project, or a combination of these, will have: requirements, designs, test plans, test cases, common libraries, documentation and coding standards, and other project artifacts. The project artifacts still have value because they represent the intellectual capital invested in the project that can be captured and made available for future projects.

Capture intellectual capital

The key to capturing the intellectual capital of a folded project is to maintain all of the project's artifacts in common storage. Typically many project artifacts are kept only on desktop systems. At some stage in the project, these artifacts are copied to the target system or to a configuration management system. This is a risky approach because information will be lost when project is stopped. If the project work is performed by contractors, they will start looking for the next consulting opportunity once the announcement is made that the project is canceled.

Don't let the contractors walk out too soon

Getting their attention and time to correctly document and save all of their work will be difficult, if even possible. In many projects once the announcement is made, all outside contractors will be off the site by the end of the day. The best rule for any project is to "capture all project artifacts in common storage and, if possible, place the artifacts in a controlled configuration management system." If you are following the approach of engineers and programmers working in a local sandbox with daily check-in of work and multiple shared versions (daily build, distributed beta, final test, and final build), then you will have all of the artifacts captured.

Reuse the knowledge

Capturing folded project artifacts allows you to reuse those artifacts in future projects.

Examples:

- Most requirements are well reviewed and "agreed to" documents and these requirements may still be valid on the next project. Many of the nonfunctional requirements, such as performance, up-time, and security, may not significantly change from project to project.
- Error handling methods
- Event logging methods
- Standard HMI faceplates, control module elements (PLC or DCS code), and interfaces
- Test strategies, frameworks, and cases
- Standard templates for requirements, code, and tests.

Each of these artifacts may have their own requirements, design documents, and code. These documents should be added to a knowledge base and made available to future project teams.

Lessons learned

Another important action to take when stopping a project is to generate a set of "lessons learned" documents. Each document should define a single aspect of the project and if it was completed above, at, or below expectations. When something worked really well, identify the participants so they can be called in on future projects to share their knowledge. When something did not work, don't identify the participants because the goal of the documents is not to assign blame. Instead, when identifying something that did not work, try to document why it didn't work, what the early symptoms were, and what could be tried instead.

Know how to fold 'em

In poker it's important to know when to hold 'em and know when to fold 'em.

In manufacturing IT projects, it is also important to know how to fold 'em.

Conclusion

The goal of this book was to educate manufacturing personnel and automation professionals in IT topics important to manufacturing. IT problems and solutions apply anywhere where software is used. Except for the special purpose languages used in automation and control systems, the same problems that plague business systems will show up in manufacturing systems, and many business system solutions can be applied to manufacturing systems. Real-time automation and control systems add an additional level of complexity in IT environments, but a good understanding of IT basics, software engineering basics, and manufacturing science informatics will help you implement robust and reliable systems.

If there is only one lesson to take away from this book, it is to "keep learning." Information technologies are ever changing and advancing, with new concepts, methods, and technologies available every year. The rate of adoption of the new technologies is also increasing. Systems and devices that start their lives as technology toys are quickly becoming mainstream and are being integrated into commercial and industrial systems. Smartphones and tablet computers went from expensive luxuries for a few to must-have devices for all in less than 5 years. With the lifetime of manufacturing systems still measured in decades, this means that there will be multiple generations of new technologies that can be integrated into your existing systems. If you keep learning, then you have the opportunity to understand the new technologies and decide why, if, and how you can incorporate them. We encourage you to use the information in this book, the books listed in the bibliography, the ISA, IEC, and ISO manufacturing IT standards, professional associations, the web, and industry magazines to keep learning and to become manufacturing IT professionals.

Bibliography

Applying S88, Batch Control From a User's Perspective, Jim Parshall, Larry Lamb, ISBN 1556177038

ANSI/ISA 99, Industrial Automation and Control Systems SecurityISA 84 and IEC 61511 Safety Integrated Systems

ANSI/ISA-84, Functional Safety: Safety Instrumented Systems for the Process Industry Sector

ANSI/ISA-88.01, Batch control—Part 1: Models and terminology

ANSI/ISA-88.02, Batch control—Part 2: Data Structures and Guidelines for Languages.

ANSI/ISA-95.01, Enterprise-Control System Integration—Part 1: Models and Terminology

ANSI/ISA-95.02, Enterprise-Control System Integration—Part 2: Object Model Attributes

Capability and Maturity Model Integration (CMMI) (http://www.sei.cmu.edu/cmmi/)

Design Patterns for Flexible Manufacturing, Dennis Brandl, ISA, ISBN 978-1-55617-998-3

IEC 61131-3, Programmable controllers—Part 3: Programming languages

IEC 61131-3, Programming Industrial Automation Systems: Concepts and Programming Languages, Requirements for Programming Systems, Decision-Making Aids, Karl Heinz John, Michael Tiegelkamp, ISBN 3642120148

IEC 61512-1, Batch control—Part 1: Models and terminology

IEC 62264-1, Enterprise/control system integration—Part 1: Models and terminology

In Pursuit of the Perfect Plan, Pat Kennedy, Evolved Technology Press, ISBN 0-9789218-6-0

Information Engineering Book 1 Introduction, James Martin, Prentice Hall, ISBN 0-13-464462-X

Information Engineering Book 2 Planning and Analysis, James Martin, Prentice Hall, ISBN 0-13-464885-4

Information Engineering Book 3 Design and Construction , James Martin, Prentice Hall, ISBN 0-13-465501-X

ISO/IEC 19501, Information technology—Open Distributed Processing—Unified Modeling Language (UML) - Version 1.4.2

ISO/IEC 19501-1 UML specification

ISO/IEC 9075-3, Information technology—Database languages—SQL—Part 3: Call-Level Interface (SQL/CLI)

Practical SIL Target Selection—Risk Analysis per the IEC 61511 Safety Lifecycle, Heidi Hartmann, Dr. Eric Scharpf, Hal Thomas, ISBN 1934977039

Heidi Hartmann (Author), Dr. Eric Scharpf (Author), Hal Thomas Safety Critical Systems Handbook: A STRAIGHTFOWARD GUIDE TO FUNCTIONAL SAFETY, IEC 61508 AND RELATED STANDARDS, INCLUDING PROCESS IEC 61511 AND MACHINERY IEC 62061 AND ISO 13849, David J. Smith and Kenneth G. L. Simpson, ISBN 0080967817

The Goal, Eliyahi M. Goldratt, North River Press, ISBN 0-88427-061-0

The Road to Integration: A Guide to Applying the ISA-95 Standard in Manufacturing, Bianca Scholten, ISA Press, ISBN 0979234387

THE WBF BOOK SERIES: *Applying ISA 88 in Discrete and Continuous Manufacturing*, Momentum Press, ISBN 160650200X

THE WBF BOOK SERIES—ISA 88 Implementation Experiences, Momentum Press, ISBN 1606502123

THE WBF BOOK SERIES: ISA 95 Implementation Experiences, Momentum Press, ISBN 1606502158

When Worlds Collide in Manufacturing Operations: ISA-95 Best Practices Book 2.0, Charlie Gifford, ISBN 1936007622

www.sei.cmu.edu/tsp in the paper "PSP: A Self-Improvement Process for Software Engineers"

XML for Dummies, Ed Tittel, IDG Books Worldwide, ISBN 0-7645-0360-X

Acronyms

ACL	Access Control List
B2MML	Business to Manufacturing Markup Language
CAPA	Corrective Action and Preventative Action
CBT	Continual Build and Test
CMM/CMMI	Capability Maturity Model/Capability Maturity Model Integration
DCS	Distributed Control System
DHCP	Dynamic Host Configuration Protocol
DLL	Dynamic Link Library
DMZ	Demilitarized Zone
DNS	Domain Name Service
ERP	Enterprise Resource Planning
ESB/EIB	Enterprise Service Bus and Enterprise Integration Bus
FAT	Final Acceptance Tests
HMI	Human Machine Interface
HTTP	Hypertext Transfer Protocol
IEC	International Electrotechnical Commission
I/O	Input/Output—Normally the digital and analog inputs and outputs of aPLC and DCS
ISA	International Society of Automation
ISO	International Organization for Standardization
IT	Information Technology

JMS	Java Messaging System
KPI	Key Performance Indicator
LIMS	Laboratory Information Management System
MDM	Master Data Management
MES	Manufacturing Execution Systems
MESA	MESA International, www.mesa.org
MI	Manufacturing Informatics OR Manufacturing Intelligence
MIB	Manufacturing Information Bus
MOM	Manufacturing Operations Management
MSI	Manufacturing Science Informatics
MULES	Mature Undocumented Legacy Execution System
NetBIOS	Network Basic Input/Output System
OAGIS	Open Applications Group Interface Standard
ODBC	Open DataBase Connection
OEE	Overall Equipment Efficiency
OPC	Name of the OPC Foundation (www.opcfoundation.org)
OPC-UA	OPC Unified Architecture
O/S	Operating System
P&ID	Piping & Instrumentation Diagram
PLC	Programmable Logic Controller
PLM	Product Life cycle Management
PSP	Personnel Software Process
RSS	Really Simple Syndication
SAML	Security Assertion Markup Language
SAT	Site Acceptance Tests
SCADA	Supervisory Control and Data Acquisition
SEI	Software Engineering Institute
Six Sigma	A process that produces 3.4 defective parts per million opportunities
SOAP	Simple Object Access Protocol
SOX	Sarbanes–Oxley Act of 2002
SPC	Statistical Process Control

SQC	Statistical Quality Control
SQL	Structured Query Language
SSO	Single Sign On
TCO	Total Cost of Ownership
TCP/IP	Transmission Control Protocol/Internet Protocol
TSP	Team Software Process
UDDI	Universal Description, Discovery, and Integration
UDP/IP	User Datagram Protocol/Internet Protocol
UML	Unified Modeling Language
UN/CEFACT	United Nations Center for Trade Facilities and Electronic Business
VLAN	Virtual LAN
WAN	Wide Area Network
WBF	Formally the World Batch Forum, now part of MESA International (www.mesa.org)
XML	Extended Markup Language
XSD	XML Schema Definition
XSLT	XSL Transformations

Index

Check Out The Other Alarm Management for Process Control Titles We Have!

Alarm Management for Process Control: A Best-Practice Guide for Design, Implementation, and Use of Industrial Alarm Systems, *Douglas H. Rothenberg*
Advanced Regulatory Control: Applications and Techniques, *David W. Spitzer*
Process Control Case Histories: An Insightful and Humorous Perspective from the Control Room, *Gregory K. McMillan*
Protecting Industrial Control Systems from Electronic Threats, *Joseph Weiss*
Industrial Resource Utilization and Productivity: Understanding the Linkages, *Anil Mital, Ph.D.*
Raw and Finished Materials: A Concise Guide to Properties and Applications, *Brian Dureu*
Going the Distance: Solids Level Measurement with Radar, *Tim Little*
Catching the Process Fieldbus: An Introduction to PROFIBUS for Process Automation, *James Powell*
Variable Speed Drives: Principles and Applications for Energy Cost Savings, Fourth Edition, *David W. Spitzer*

Announcing Digital Content Crafted by Librarians

Momentum Press offers digital content as authoritative treatments of advanced engineering topics, by leaders in their fields. Hosted on ebrary, MP provides practitioners, researchers, faculty and students in engineering, science and industry with innovative electronic content in sensors and controls engineering, advanced energy engineering, manufacturing, and materials science. **Momentum Press offers library-friendly terms:**

- perpetual access for a one-time fee
- no subscriptions or access fees required
- unlimited concurrent usage permitted
- downloadable PDFs provided
- free MARC records included
- free trials

The **Momentum Press** digital library is very affordable, with no obligation to buy in future years.

For more information, please visit **www.momentumpress.net/library** or to set up a trial in the US, please contact **Adam Chesler**, *adam.chesler@momentumpress.net*.

www.ingramcontent.com/pod-product-compliance
Lightning Source LLC
LaVergne TN
LVHW062125200926
848484LV00007B/1768
* 9 7 8 1 6 0 6 5 0 4 3 1 4 *